Springer Proceedings in Complexity

Springer Proceedings in Complexity publishes proceedings from scholarly meetings on all topics relating to the interdisciplinary studies of complex systems science. Springer welcomes book ideas from authors. The series is indexed in Scopus.

Proposals must include the following:

- name, place and date of the scientific meeting
- a link to the committees (local organization, international advisors etc.)
- scientific description of the meeting
- list of invited/plenary speakers
- an estimate of the planned proceedings book parameters (number of pages/articles, requested number of bulk copies, submission deadline)

Submit your proposals to: Hisako.Niko@springer.com

Mariana Macedo · Alessio Cardillo ·
Wellington Franco · Angelo Brayner ·
Ronaldo Menezes
Editors

Complex Networks XVI

Proceedings of the 16th Conference on Complex Networks, CompleNet 2025

Springer

Editors
Mariana Macedo
Department of Data Science
Northeastern University London
London, UK

Wellington Franco
Universidade Federal do Ceará
Fortaleza, Brazil

Ronaldo Menezes
Department of Computer Science
University of Exeter
Exeter, UK

Alessio Cardillo
Departament de Física de la Materia
Condensada
Universitat de Barcelona
Barcelona, Spain

Angelo Brayner
Universidade Federal do Ceará
Fortaleza, Brazil

ISSN 2213-8684 ISSN 2213-8692 (electronic)
Springer Proceedings in Complexity
ISBN 978-3-031-93618-0 ISBN 978-3-031-93619-7 (eBook)
https://doi.org/10.1007/978-3-031-93619-7

This Springer imprint is published by the registered company Springer Nature Switzerland AG
The registered company address is: Gewerbestrasse 11, 6330 Cham, Switzerland

Message from the Programme Chairs

The International Conference on Complex Networks—CompleNet was proposed in 2008, and the first workshop took place in 2009 in Catania (Italy). The initiative was led by Ronaldo Menezes (Department of Computer Science, University of Exeter, UK, formerly at the BioComplex Laboratory, Florida Institute of Technology, USA) and Giuseppe Mangioni (Dipartimento di Ingegneria Elettrica, Elettronica e Informatica, Università di Catania, Italy). CompleNet brings together researchers from various areas related to Complex Networks (Network Science), offering both a platform for full paper publications in the conference proceedings and an opportunity to present abstracts of ongoing research. This dual format makes the conference appealing to researchers seeking peer-reviewed publication outlets as well as those wanting to share and discuss their current work published in other venues (or ongoing). From biology to urban systems, from economics to social systems, complex networks are becoming pervasive in many fields of science. It is the interdisciplinary nature of complex networks that CompleNet aims to grasp. CompleNet 2025 is the 16th event in the series, and has been hosted by the Federal University of Ceará at a historical venue, Casa José Alencar in Fortaleza, from April 22nd to April 25th, 2025.

The present book includes the peer-reviewed list of works presented at CompleNet 2025. We received 111 submissions from 16 countries. Each submission was reviewed by at least three members of the Programme Committee. Acceptance was determined on the relevance to the symposium themes, clarity of presentation, originality and accuracy of results and proposed solutions. After the review process, 11 papers were selected for inclusion in this book. These contributions belong to several topics related to complex networks such as *Computational Social Science, Dynamics on and of Networks, Ecological Networks, Epidemic Modelling in Networks, Network Algorithms, Network Evolution and Growth, Applications of Networks, Network Theory and Models*, and *Networked Medicine.*

In the following, we provide a brief summary of each contribution appearing in this book in order of appearance. Morais and Interian (Chapter "A Simple and Flexible Algorithm to Generate Real-World Networks") propose a novel algorithm for generating networks with suitable average distances and clustering coefficients. In particular, the algorithm leverages only local information on the network's topology.

Trindade, Dreveton, and Figueiredo (Chapter "A Framework for Efficient Estimation of Closeness Centrality and Eccentricity in Large Networks") introduce a new framework for calculating node closeness and eccentricity in large networks using an anchor-based approximation. Silva, Santana, Menezes, and Bastos-Filho (Chapter "Network-Centric Analysis of Memetic Operators and Communication Topologies for Swarm Intelligence Algorithms") apply network science to compare various Particle Swarm Optimization (PSO) variants, demonstrating that memetic operators and communication topologies play a crucial role in the PSO performance. Silva, Silva, Bastos-Filho, Rosa, Albuquerque, Rodrigues, Tahara, Roisman and Moscavitch (Chapter "A Hybrid Framework for Quantifying and Analyzing the Structural Properties of Human Retinal Vessel Networks") apply Network Science to analyse the structure of retinal vessel networks. In particular, by mapping each network into an array (whose components are the network's topological descriptors) the authors cluster networks trying to identify potential biomarkers for diseases such as diabetes and hypertension. Costa, Ribeiro, and Neto (Chapter "An Approach Based on Networks and Machine Learning for Gastric Cancer Treatment Recommendation") leverage network science and machine learning to extract insights into gastric cancer, contributing to improved treatment strategies by developing a precision medicine/ treatment approach. Moreover, the algorithm is tested on real case scenarios from gastric oncology clinical practice in Brazil. Cavalcante Basílio and Figueiredo (Chapter "Population Dynamics in the Global Coral–Dymbiont Network Under Temperature Variations") present an ecological model of population dynamics occurring between corals and zooxanthellae algae (symbiont) species accounting for the empirical structure of the coral–symbiont bipartite network, whose features are sensible to recurrent warming events. The results reveal that the nodes' degree plays a significant role in population growth after successive warming events, with generalist species displaying higher levels of growth across all ocean regions analysed. Lima and Atman (Chapter "Dengue Serotypes Cyclicity Evidenced by the Impact-Frequency Histogram of the Visibility Graph") apply the visibility graph approach to convert time-series of dengue disease incidence into networks. They consider data of two Brazilian cities: Rio de Janeiro and Belo Horizonte. They introduce an impact-frequency (a proxy for the incidence's variation over time) histogram protocol to evaluate cyclic dengue patterns. Then, they analyse the empirical time series of incidence and estimate from them the period for dengue's re-infestation. Santos, Pereira, Murari, Filho, and Cunha (Chapter "The Brazilian Maritime Network During the COVID-19 Pandemic: Analysis of Topologies and Impacts on Connectivity") examine the variations of the topological features of the Brazilian maritime network during the COVID-19 pandemic by leveraging data from the Automatic Identification System, which monitors the movements of ships between Brazilian ports. The results highlight the strengthening of regional hubs (e.g., Manaus and Suape), a redistribution of cargo flows, a decrease of the network's modularity, and an increase of the average path length (i.e., higher time and costs). Taveira, Buarque de Lima Neto, and Menezes (Chapter "Understanding the Structure and Resilience of the Brazilian Federal Road Network Through Network Science") map the Brazilian federal road network into a set of weighted networks. They consider different types of weights like distance, number of accidents, populations' sizes, and then perform a topological characterization of each network. They use such analysis to extract central nodes (i.e., cities), as well as clusters of cities corresponding to the network's communities. Moinard and Latapy (Chapter "Improving Flocking Behaviors in Street Networks with Vision") extend their previous flocking model of walkers on a road network by expanding the walkers' field of vision. Such an expansion guarantees that walkers do not split anymore into divergent directions when they arrive at road intersections. This phenomenon allows to simulate groups of walkers whose gathering times and robustness to break ups are more realistic. Bakker and Rodriguez-Rivero (Chapter "Social Circles Impact on Opinion Dynamics") evaluate the impact of *social circles* (i.e., groups of people sharing some attribute) on the emergence of social consensus in online social networks. The phenomenology observed points out that adding social circle information to opinion dynamics models results in a lower level of consensus. By comparing the mean level of consensus per social circle and per community the authors observe that the correlation between the final distribution of opinions within social circles is higher than within communities.

We would like to thank the Programme Committee members sprawled across 21 countries for their work in promoting the event and acting as quality gatekeepers by refereeing submissions. In particular, we want to acknowledge the work of:

Albert Diaz-Guilera—Universitat de Barcelona (Spain)
Alberto Antonioni—Carlos III University of Madrid (Spain)
Alexandre Evsukoff—Universidade Federal do Rio de Janeiro (Brazil)
Andreia Sofia Teixeira—Northeastern University London (UK)
Angélica da Mata—Universidade Federal de Lavras (Brazil)
Anna Lawniczak—University of Guelph (Canada)
Anthony Perez—LIFO, Université d'Orléans (France)
Attila Szolnoki—Centre for Energy Research (Hungary)
Avner Bar-Hen—CNAM, Paris (France)
Carlo Piccardi—Politecnico di Milano (Italy)
Carmelo Bastos Filho—University of Pernambuco (Brazil)
Claudio Castellano—Istituto dei Sistemi Complessi (ISC-CNR) (Italy)
Dan Braha—New England Complex Systems Institute (USA)
David Soriano-Paños—Universitat Rovira i Virgili (Spain)
Diego Pinheiro—Universidade de Pernambuco (Brazil)
Douglas Ferreira—Federal Institute of Rio de Janeiro (Brazil)
Elisa Omodei—Central European University (Austria)
Esteban Moro—Northeastern University (USA)
Eszter Bokanyi—University of Amsterdam (The Netherlands)
Eytan Katzav—Hebrew University of Jerusalem (Israel)
Fakhteh Ghanbarnejad—SRH Berlin University of Applied Sciences (Germany)
Filippo Radicchi—Indiana University (USA)
Francesco Cauteruccio—University of Salerno (Italy)
Francisca Ortiz—Millennium Institute for Care Research (MICARE) (Chile)

Frank Takes—Leiden University (The Netherlands)
Gareth Baxter—University of Aveiro (Portugal)
Gergely Palla—Health Services Management Training Centre, Semmelweis University (Hungary)
Giacomo Livan—University of Pavia (Italy)
Giulia Cencetti—CNRS (France)
Giulio Rossetti—CNR-ISTI (Italy)
Giuseppe Mangioni—University of Catania (Italy)
Hernan Makse—City College of New York (USA)
Hocine Cherifi—University of Burgundy (France)
Hugo Pérez-Martínez—Universidad de Zaragoza (Spain)
Ivana Bachmann—Universidad de Chile (Chile)
Javier Borge-Holthoefer—Universitat Oberta de Catalunya (Spain)
Jesús Gómez Gardeñes—Universidad de Zaragoza (Spain)
Johannes Wachs—Corvinus University of Budapest (Hungary)
Jordi Duch—Universitat Rovira i Virgili (Spain)
Jordi Nin—BBVA Data and Analytics (Spain)
José Mendes—Universidade de Aveiro (Portugal)
Kwang-Il Goh—Korea University (South Korea)
Letizia Milli—University of Pisa (Italy)
Lorenzo Zino—Politecnico di Torino (Italy)
Lucila Alvarez Zuzek—Fondazione Bruno Kessler (Italy)
Manlio De Domenico—University of Padua (Italy)
Matthieu Latapy—CNRS (France)
Michael Szell—IT University of Copenhagen (Denmark)
Michele Coscia—IT University of Copenhagen (Denmark)
Mincheng Wu—Zhejiang University of Technology (China)
Osamu Sakai—University of Shiga Prefecture (Japan)
Pablo Balenzuela—University of Buenos Aires (Argentina)
Per Sebastian Skardal—Trinity College (USA)
Peter Pollner—ELTE (Hungary)
Rafael Prieto-Curiel—Complexity Science Hub (Austria)
Riccardo Di Clemente—Northeastern University London (UK)
Richard Tillquist—California State University, Chico (USA)
Robert Benassai—Universitat Oberta de Catalunya (Spain)
Rubén Rodríguez-Casañ—Universitat Oberta de Catalunya (Spain)
Sadamori Kojaku—Binghamton University (USA)
Sandro Meloni—IFISC—CSIC (Spain)
Sanjukta Bhowmick—University of North Texas (USA)
Satyam Mukherjee—Shiv Nadar University (India)
Sergey Shvydun—Delft University of Technology (The Netherlands)
Sergi Lozano—Universitat de Barcelona (Spain)
Sergio Gómez—Universitat Rovira i Virgili (Spain)
Stephany Rajeh—EFREI Paris-Panthéon-Assas University (France)
Stephen Uzzo—National Museum of Mathematics (USA)

Thilo Gross—HIFMB Helmholtz Institute for Functional Marine Biodiversity (Germany)
Tim Evans—Imperial College London (UK)
Timoteo Carletti—University of Namur (Belgium)
Tiziano Squartini—IMT School for Advanced Studies Lucca (Italy)
Violeta Calleja Solanas—Doñana Biological Station—CSIC (Spain)
Yifang MA—Southern University of Science and Technology (China)

Finally, we are grateful to our keynote speakers: José S. Andrade Jr., Celia Anteneodo, Carmen Cabrera-Arnau, Philipp Lorenz-Spreen, Esteban Moro, and Francisca Ortiz-Ruiz. Their presentations were one of the reasons that made CompleNet 2025 a resounding success.

Alessio Cardillo
Mariana Macedo
Programme Chairs

Message from the Conference Chairs

As we celebrated the 16th anniversary of the CompleNet Conference, the event was proudly hosted by the Federal University of Ceará in Fortaleza, Brazil. The 2025 proceedings represent our long-standing commitment to quality, while being a venue encouraging participation all in a plenary-only environment. CompleNet continues in the evolving landscape of Network Science. This unique edition distinguished itself with a commitment to delivering comprehensive insights into the field's latest research and innovations in an engaging, multidisciplinary format that eschewed parallel sessions.

The 2025 edition, held at the historic Casa José de Alencar in the vibrant coastal city of Fortaleza, Brazil, provided participants with more than just a forum for cutting-edge Network Science. The venue itself, a cultural heritage site and former home of one of Brazil's most celebrated nineteenth-century novelists, José de Alencar, offered a unique blend of historical significance and academic atmosphere. It also offered an immersive cultural and natural experience. Attendees had the opportunity to explore Fortaleza's beautiful coastline and historic landmarks, as well as savouring the local cuisine and outdoor activities. This blend of academic excellence and local allure emphasised CompleNet's dedication to fostering a holistic experience for the international community.

Network Science is a field intersecting with Data Science and Complex Systems, focusing on the study of complex networks such as social networks, technological networks, and biological networks. It is pivotal in understanding the structure and dynamics of interactions within these systems. The importance of Network Science lies in its ability to uncover patterns, identify influential components, and predict system behaviours. Its connection to complex systems provides valuable insights into emergent properties, and helps organisations make more informed decisions by understanding how different components influence each other. Its relation to Data Science is symbiotic; while Network Science provides the framework and principles for understanding the connections, Data Science offers the tools and methodologies for analysing and interpreting the vast amounts of data these networks generate. Together, they enable a deeper comprehension of complex systems in nature and society, leading to innovations in various sectors including healthcare, technology, and urban planning.

The International Conference on Complex Networks (CompleNet) stands as a pivotal gathering that has united researchers and practitioners from diverse fields, all focusing on complex networks, since its inception in 2009. The interdisciplinary approach of CompleNet has been illuminating the widespread application of complex networks across biological, technical, informational, economic, and social systems. Esteemed for its plenary sessions and the balance between young and experienced researchers, CompleNet 2025 invited participants to delve into the latest Network Science developments.

The contributions within these proceedings reflect the extensive range of topics addressed at the conference, such as Computational Social Science, Dynamics *on* and *of* Networks, Ecological Networks, Epidemic Modelling in Networks, Network Algorithms, Network Evolution and Growth, Applications of Networks, Network Theory and Models, and Networked Medicine. The papers included in this volume certainly advances our understanding and the development of Network Science as a whole.

In addition to these papers, the conference also showcased a rich program of abstracts. In a significant departure from previous years, we made the innovative decision to offer all participants the opportunity to present their work in oral, plenary sessions, fostering greater engagement and visibility across the community. Though not included in this publication, these abstracts were pivotal in fostering dynamic discussions and highlighting emerging research directions within our community. This year's conference not only underscored the vibrancy and diversity of Network Science but also reinforced its role in addressing complex challenges across disciplines. The enthusiastic participation and high-quality submissions confirmed the ongoing relevance and impact of the field.

A highlight of CompleNet 2025 was the thought-provoking panel discussion chaired by Fernando Buarque de Lima Neto, which explored the critical intersection between Network Science and Responsible Artificial Intelligence. This timely session underscored the growing importance of ethical considerations in technological advancement, particularly as network-based AI systems become increasingly prevalent in society. The discussion illuminated how Network Science's frameworks for understanding complex interactions can inform the development of more transparent and accountable AI systems while simultaneously highlighting how responsible AI practices can enhance our ability to analyse and interpret complex networks. The synergy between these fields proves essential as we face the challenges of developing AI systems that are not only technically sophisticated but also aligned with human values and societal needs.

In a groundbreaking initiative exemplifying CompleNet's commitment to nurturing emerging talent and promoting inclusivity in science, the 2025 edition featured our first-ever dedicated session for students from the state of Ceará. This special track invited only local students to submit abstracts for evaluation, providing them with an invaluable opportunity to participate in an international conference early in their academic careers. This was offered to them free of charge, with all the costs being covered by the event. The initiative served as a bridge between local academic communities and the global scientific discourse, offering these young researchers not only exposure to cutting-edge research, but also the chance to network with established scientists from around the world. This program aligns perfectly with our mission to democratise access to scientific knowledge and foster the development of the next generation of network scientists, particularly in the Global South.

CompleNet takes immense pride in bringing Network Science to the Global South, recognising that true scientific progress can only be achieved through genuine global participation and diverse perspectives. While hosting international conferences outside the traditional Europe–USA axis presents unique challenges, we firmly believe that these efforts are fundamental to fostering a more inclusive and equitable scientific community. The rich exchanges and collaborations emerging from such geographical diversity benefit not only the local academic communities but also enhance the global scientific discourse through the integration of diverse viewpoints and experiences.

In an era where some forces seek to undermine progress in equality, diversity, and inclusion, CompleNet stands resolute in its commitment to these fundamental values. This commitment is reflected not only in our choice of conference locations over the years but also in the careful consideration we give to representing diverse voices among our keynote speakers—spanning different career stages, genders, and institutional affiliations across the globe. We believe that this diversity strengthens our scientific community and leads to more innovative and comprehensive approaches to understanding complex networks.

Our heartfelt gratitude goes to the Programme Chairs, Mariana Macedo (Northeastern University London, UK) and Alessio Cardillo (University of Barcelona, Spain), as well as the members of the programme committee they assembled. Their commitment to promoting the event and evaluating submissions was commendable.

We are particularly indebted to our Local Organisation Chair, Wellington Franco (Federal University of Ceará), whose tireless dedication and exceptional organisational skills were instrumental to the success of this conference. His profound understanding of both the local context and the academic requirements ensured that every aspect of the conference ran smoothly. We can confidently say that without his remarkable efforts and commitment, this conference would not have achieved the level of excellence it did.

We extend special thanks to Célio Sousa (@celio.f.sousa on Instagram), the talented local sculptor whose commissioned artworks served as memorable conference mementos and prizes for attendees. His unique artistic vision helped create lasting memories of CompleNet 2025 and Fortaleza's rich cultural heritage.

We are also honoured by the participation of our invited speakers, listed here in alphabetical order, for their enriching discussions and perspectives: José S. Andrade Jr. (Federal University of Ceará, Brazil), Celia Anteneodo (PUC-RJ, Brazil) Carmen Cabrera-Arnau (University of Liverpool, UK), Philipp Lorenz-Spreen (Max Planck Institute and TU Dresden, Germany), Esteban Moro (Northeastern University, USA), and Francisca Ortiz-Ruiz (Universidad Mayor, Chile). Their presentations were instrumental in the success of this year's conference, highlighting the rich diversity and depth of Network Science research.

Lastly, we extend our profound gratitude to our sponsors and partners, whose support has been invaluable to the success of this event. We are particularly grateful to Springer Nature and the Applied Network Science journal for their longstanding and continued support of CompleNet over the years. We also thank the following journals: Entropy (MDPI), Computer Science (PeerJ Publishing), and JPhys Complexity (IOP Publishing), for their valuable sponsorship of this year's conference. Their commitment to scientific publishing and the advancement of Network Science has significantly contributed to the success of our event.

Fortaleza, Brazil
April 2025

Ronaldo Menezes
Angelo Brayner
Conference Chairs

Contents

A Simple and Flexible Algorithm to Generate Real-World Networks ... 1
João Pedro C. Morais and Ruben Interian

A Framework for Efficient Estimation of Closeness Centrality and Eccentricity in Large Networks ... 13
Patrick C. Trindade, Maximilien Dreveton, and Daniel R. Figueiredo

Network-Centric Analysis of Memetic Operators and Communication Topologies for Swarm Intelligence Algorithms ... 27
Carlos Silva, Clodomir Santana, Ronaldo Menezes, and Carmelo Bastos-Filho

A Hybrid Framework for Quantifying and Analyzing the Structural Properties of Human Retinal Vessel Networks ... 41
Hitalo Silva, Diego Silva, Carmelo Bastos-Filho, Alexandre Rosa, Rafael Albuquerque, Arlington Rodrigues, Luigi Tahara, Luiz Roisman, and Samuel Moscavitch

An Approach Based on Networks and Machine Learning for Gastric Cancer Treatment Recommendation ... 55
Lucas Queiroz Melo da Costa, Carlos Henrique Costa Ribeiro, and Emmanuel Dias-Neto

Population Dynamics in the Global Coral-Symbiont Network Under Temperature Variations ... 69
Maria Gabriella Cavalcante Basílio and Daniel Ratton Figueiredo

Dengue Serotypes Cyclicity Evidenced by the Impact-Frequency Histogram of the Visibility Graph ... 83
L. L. Lima and A. P. F. Atman

The Brazilian Maritime Network During the COVID-19 Pandemic: Analysis of Topologies and Impacts on Connectivity 95
Carlos César Ribeiro Santos, Hernane Borges de Barros Pereira, Thiago Barros Murari, Leonardo Sanches de Carvalho Filho, and Marcelo do Vale Cunha

Understanding the Structure and Resilience of the Brazilian Federal Road Network Through Network Science 107
Júlio Taveira, Fernando Buarque, and Ronaldo Menezes

Improving Flocking Behaviors in Street Networks with Vision 123
Guillaume Moinard and Matthieu Latapy

Social Circles Impact on Opinion Dynamics 135
Emy B. M. Bakker and Cristian Rodriguez Rivero

A Simple and Flexible Algorithm to Generate Real-World Networks

João Pedro C. Morais and Ruben Interian

Abstract This study introduces an algorithm that generates undirected graphs with three main characteristics of real-world networks: scale-freeness, short distances between nodes (small-world phenomenon), and large clustering coefficients. The main idea is to perform random walks across the network and, at each iteration, add special edges with a decreasing probability to link more distant nodes, following a specific probability distribution. A key advantage of our algorithm is its simplicity and flexibility in creating networks with different characteristics without using global information about network topology. We show how the parameters can be adjusted to generate networks with specific average distances and clustering coefficients, maintaining a long-tailed degree distribution. The implementation of our algorithm is publicly available on a GitHub repository.

Keywords Real-world network · Scale-free network · Random walk · Small-world phenomenon · Clustering coefficient

1 Introduction

In recent years, the study of complex networks has gained prominence across various scientific disciplines, such as sociology, physics, biology, and computer science [1–5]. Real-world networks often display both scale-free characteristics and surprising proximity among network nodes. These two features commonly observed in real-world networks are the so-called Matthew effect and the small-world phenomenon.

The Matthew effect (also known as "rich-get-richer effect", or accumulated advantage) reflects a preferential attachment dynamic [6], leading to the emergence of high-degree hubs in a network and long-tailed degree distributions [7]. This behavior can be easily observed in social networks, where most people have few connections, and few people hold a very large number of connections, showing a long-tail pattern [8].

J. P. C. Morais · R. Interian (✉)
Instituto de Computação, Universidade Estadual de Campinas (UNICAMP), São Paulo, Brazil
e-mail: ruben@ic.unicamp.br

M. Macedo et al. (eds.), *Complex Networks XVI*, Springer Proceedings in Complexity,
https://doi.org/10.1007/978-3-031-93619-7_1

Social networks are also examples of networks with high clustering, where individuals tend to form tightly connected groups, such as friend circles, family groups, or professional communities. In these clusters, the likelihood of any two friends of some individual also being friends is higher than in random networks, resulting in a high clustering coefficient.

Finally, the small-world property characterizes networks with short average path lengths, enhancing navigability. Collaboration networks of actors and the neural network of the nematode worm *C. elegans* are examples of small-world networks [9].

Modeling networks with these three characteristics can be challenging, as the mechanisms underlying each of them tend to drive network structures in different directions. Random walk models [10] were previously used to generate networks with long-tailed degree distributions, capturing real-world systems' natural growth and preferential attachment. A key advantage is their ability to create structures without global information, aligning with the organic development of real networks. While effective at generating hubs and modeling the Matthew effect, traditional random walks often overlook reducing path lengths between nodes—a crucial feature for small-world networks. This limitation makes it challenging to model systems requiring long-tailed degree distributions, high clustering, and short average distances.

In this paper, we present a random walk algorithm designed to generate networks that show small-world features, Matthew effect, and high local clustering, adding special edges that emulate randomness in link generation observed in real networks, thus reducing the average distances between the nodes in the network.

The paper is organized as follows. Section 2 presents previous works that inspired and contributed ideas that led to the presented approach. In Sect. 3, we describe our algorithm. The results and characteristics of the generated networks are presented in Sect. 4. Conclusions are detailed in the last section.

2 Previous Works

Previous works have studied the generation of networks with specific combinations of features considered in our study.

Arguing that real-world networks are often highly clustered while showing small average distances between nodes, Watts and Strogatz [9] proposed a model to reproduce such characteristics. Starting with a set of N nodes in a circular order where each node is connected with an undirected edge to k neighbors, the authors rewired each edge with some fixed and small probability p. The average clustering coefficient remained quite high while the average distance dropped to a small value approximately proportional to $\log N$.

Latent-space network models have also been employed to investigate small-world networks with non-vanishing clustering [11]. In its simplest version, also called random geometric graph model, nodes are distributed uniformly at random in some metric space, and two nodes i and j are connected if and only if the distance x_{ij}

between them is less than some parameter μ, leading to high clustering and large average shortest paths. However, the model can become small-world by introducing a probability p_{ij} of the existence of a link between the nodes. For example, in $\mathbb{R}^d$ space, choosing $p_{ij} \propto x_{ij}^{-\beta}$ with $\beta \in (d, 2d)$ results in non-vanishing clustering coefficients and small-world networks, with average distances scaling proportionally to $\log N$ [12].

On the other hand, real-world networks also have high-degree nodes called hubs, which are absent in the above models. Barabási and Albert [6] proposed a preferential attachment process that generates such long-tailed degree distributions. Starting from some small graph, one new node v is added at each iteration with k new edges linking v to k different nodes chosen with probabilities proportional to node degrees. That is, the likelihood of node v choosing node w is proportional to the degree of w at that iteration, generating scale-free networks.

To increase the clustering coefficient of the networks generated by the above BA model, Holme and Kim [13] introduced a Triad Formation step performed with probability P_t after a node and its edges are added to the network. When a new edge is added linking the new node v to a node w, an edge is added linking v to a randomly selected neighbor of w, thus creating a triad between the three nodes and increasing the clustering of the network.

Although previous models have presented solid results in scale-free network construction, a significant concern we raise is that they require global information at each step (e.g., the degrees of all nodes to calculate the preferential attachment probabilities), which may be unrealistic since in real-world networks links emerge naturally, without knowing global information about network topology.

Saramäki and Kaski [10] proposed using random walkers to generate undirected scale-free networks, showing that it is not necessary to have global information about node degrees at each step to achieve such results. Herrera and Zufiria [14] improved this process by using the number of steps in the random walks to guide triangle generation, introducing a way to control the network's clustering coefficient using again only local information.

The random walk process proposed by Saramäki and Kaski [10] begins from a typically small initial graph with n_0 nodes. At each iteration, a new node v is added to the graph, linking v to existing nodes that will be chosen using random walks. These chosen nodes (called "marked nodes" from now on) are identified as follows: beginning from a randomly selected node w, l random steps are taken from w, allowing to revisit previous nodes. After each walk, the endpoint is marked, and the process continues until m nodes are marked. The new node v is then connected to marked nodes. The process finalizes after adding N new nodes to the graph.

Herrera and Zufiria [14] noted that by changing the value of the parameter l, the number of steps in the random walk, it is possible to control the network's clustering coefficient. If $l = 1$, the neighbor of a marked node will also be marked, generating a triangle between these two neighbors and the added node, thus affecting the clustering coefficient. Each node v has an associated value p_v, the probability of $l = 1$ if the random walk starts from that node.

Random walks controlling the walk length l proved an efficient way to create power-law networks while regulating the clustering coefficient. However, the lack of attention to the average distances on the network remained an open question for real-world network generation using random walks.

3 Methods

The primary goal of our algorithm is to generate long-tailed networks that combine high clustering coefficients with short path lengths between the nodes.

Our algorithm starts from a small initial graph $G(V, E)$ with n_0 nodes. A new node v is added to the graph at each iteration. Following Saramäki and Kaski [14], we perform a random walk, starting from a random node. We mark this initial node and then continue to mark each node reached after l steps along the random walk, resulting in m marked nodes. Edges from v to each of the m marked nodes are added to the graph at the end of the walk.

To control the clustering coefficient, it can be used l, the number of steps between any two nodes marked successively. The value of l is decided after marking some node, and before starting a new phase of the random walk, having some probability p_1 to be 1, and probability $1 - p_1$ to be 2.

Note that l follows a Bernoulli-like distribution, as proposed in [14], but it is possible to modify this distribution to contemplate larger values for l. However, as we will see later, we can already achieve the desired behavior with this simple distribution of l.

At this point, unlike in [10, 14], an additional edge is created at each iteration. The idea behind this step is to reduce the overall distances within the network (a neglected aspect in the original model) by adding shortcut edges, but not in an entirely random manner. The process begins by choosing a value d with some probability $P(d)$. The distribution $P(d)$ is fixed for each algorithm execution, and decreases as d grows in such a way that larger values of d are less likely to be chosen than smaller ones. A random node s in the network is then chosen, and we pick another node t located at d steps from s. The new edge then connects s and t.

The probability distribution we used is based on the idea that the likelihood of linking two nodes, x and y, should decrease inversely proportional to the distance between x and y squared. In this way, shortcuts are created on the network, but not in a completely random manner, but somewhat closer to the link generation process in real networks, where more similar nodes are more likely to be connected. For example, two individuals who are closer to each other, in the geographical or topological sense, are more likely to meet. The emergence of these 'random' edges reflects the natural appearance of new relations and links between nodes as the network grows.

Thus, in our model $P(d) \propto \frac{1}{d^2}$, or $P(d) = \frac{A}{d^2}$ for some fixed A, since the probabilities for each d value we use should sum to 1. The value of the normalizing constant A may be found using the fact that

$$A\left(\frac{1}{2^2} + \frac{1}{3^2} + \cdots + \frac{1}{d_{max}^2}\right) = 1,$$

where d_{max} is the approximate current diameter of the graph. We estimate d_{max}, an approximation to the real diameter, in a simple way: assume that each node has a degree equal to the average degree $\overline{deg} = \frac{2|E|}{|V|}$ over all nodes. In an exponential branching process, at each distance k from some node v, there are approximately $(\overline{deg} - 1)^k$ nodes. Using the geometric series sum, there are at all $\frac{(\overline{deg}-1)^k-1}{\overline{deg}-2}$ nodes at a distance at most k from v. Growing k, at some point, the number of covered nodes will reach the overall number of nodes $|V|$, being:

$$\frac{(\overline{deg} - 1)^k - 1}{\overline{deg} - 2} = |V|$$

$$(\overline{deg} - 1)^k = |V| \cdot (\overline{deg} - 2) + 1$$

$$k = \log_{\overline{deg}-1} \left(|V| \cdot (\overline{deg} - 2) + 1\right)$$

The sought diameter d_{max} is then $2 \log_{\overline{deg}-1} \left(|V| \cdot (\overline{deg} - 2) + 1\right)$, twice the maximum value of k. The chosen value for d is unlikely to be bigger than the real diameter of the network.

In a simplified way, the proposed algorithm that builds the network goes as follows:

1. Start with a small initial graph with n_0 nodes.
2. Add a new node v to the network.
3. Pick a random node w. Perform a random walk from w, marking each node reached after every l steps. Stop when m nodes are marked, and connect them to v.
4. Choose a value d with some probability $P(d)$ and a random node s. Find a node t at a distance d from s, and connect s and t with an edge.
5. Repeat N times the steps 2–4.

Algorithms 1 and 2 describe in more detail our implementation of the network generation approach presented in this paper. Algorithm 1 illustrates the random walk process that starts from a node *start* in a graph G, uses the probability p_1 of l being equal to 1, and generates a list of m marked nodes.

In line 1, the RandomWalk algorithm initializes the marked list with the starting node, and in line 2, it sets the current node that tracks the position during the walk to *start*. Lines 3–8 are repeated $m - 1$ times, adding $m - 1$ nodes to the list of marked nodes. In line 4, the value of l is established based on the parameter p_1, and in lines 5–7, the new phase of the random walk is performed. The endpoint is then added to the marked node list in line 8. Line 9 returns the list of marked nodes.

Algorithm 1 RandomWalk $(G, start, p_1, m)$

marked ← {start}
current ← start
repeat $m - 1$ **times**
    l ← takes the value 1 with probability p_1 and 2 otherwise
    repeat l **times**
        neighbors ← G[current].neighbors
        current ← neighbors[randomIndex]
    marked ← marked ∪ {current}
return marked

On the other hand, Algorithm 2 describes the introduced network generation process. It takes as parameters an initial graph G, which is typically small, the number of nodes N to be added to the graph, the probability p_1 of l being equal to 1, and the number of edges m to be added at each iteration.

The algorithm repeats N times the following sequence of steps. It chooses a random node of the network as the starting point of the random walk in line 2. Then, the random walk is performed in line 3 by the procedure RandomWalk, returning the list of marked nodes. In line 4, a new node v is added to the network, and in line 5, the edges are created between v and the marked nodes. Lines 6–9 describe the process of adding an extra edge by choosing a distance d based on some probabilistic distribution, and creating an edge between a random node s and a node t at d steps from s. Line 10 returns the resulting network.

Algorithm 2 GenerateNetwork (G, N, p_1, m)

1: **repeat** N **times**
2:     start ← random node of the network
3:     marked ← **RandomWalk**$(G, start, p_1, m)$
4:     v ← G.**add_node**()
5:     **for** $u \in$ marked **do** G.**add_edge**(v, u)
6:     d ← random value according to distribution $P(d)$
7:     s ← random node of the network
8:     t ← **find_node**(s, d)
9:     G.**add_edge**(s, t)
10: **return** G

4 Results

This section presents the main characteristics of the networks generated by the proposed algorithm. We show how it can be used to create graphs with characteristics similar to those found in real-world networks. In our experiments, we used as the initial graph G a cycle with 10 nodes (circular graph, C_{10}). Each row in a table represents one execution of our algorithm.

Our analysis focuses mainly on four key network measures: the average local clustering coefficient $\overline{C}$, which indicates the likelihood that two neighbors of a given node are also neighbors, averaged across all nodes in the network; global clustering coefficient C (or transitivity), which is the ratio of closed triplets to the total number of triplets in the graph; the average shortest path length ($\overline{L}$), representing the mean shortest distance between any pair of nodes; the estimated power-law coefficient, denoted by γ, which characterizes the degree distribution of the network in the following way: the probability $P(k)$ that a randomly selected node has degree k is approximately proportional to $k^{-\gamma}$. The exponent is approximated by calculating the angular coefficient of the degree distribution on a log-log scale using the least squares method.

Our results show that the algorithm can generate a wide range of different networks. Table 1 shows that the clustering coefficients grow proportionally to the value of p_1, reaching a fairly high value for $p_1 = 1$, without affecting the other measures. This behavior can also be seen in Fig. 1. Thus, by changing p_1 with a fixed m, it is possible to generate networks with different clustering coefficients, providing a simple way to control the value of this measure in the generated graphs. From Table 1, we can also see that γ, the power-law exponent, varies little with p_1, staying almost constant after $p_1 = 0.3$.

Table 1 Generated networks and their measures using different p_1 values ($m = 5$, $N = 50000$). Parameters: probability p_1 of l being equal to 1. Measures: average local clustering coefficient $\overline{C}$; transitivity C; average shortest path length $\overline{L}$; estimated power-law exponent γ; maximum degree of the network d_{max}

p_1	$\overline{C}$	C	$\overline{L}$	γ	d_{max}
0.0	0.0461	0.0193	4.2705	−1.9766	525
0.1	0.0848	0.0310	4.2813	−2.2508	382
0.2	0.1207	0.0412	4.2889	−1.9858	426
0.3	0.1512	0.0501	4.2918	−2.2084	448
0.4	0.1814	0.0594	4.3115	−2.2346	449
0.5	0.2104	0.0680	4.3351	−2.2049	417
0.6	0.2390	0.0760	4.3494	−2.1996	396
0.7	0.2680	0.0841	4.3731	−2.2039	546
0.8	0.2971	0.0935	4.4125	−2.1980	427
0.9	0.3259	0.1033	4.4498	−2.2459	403
1.0	0.3549	0.1108	4.4762	−2.2615	440

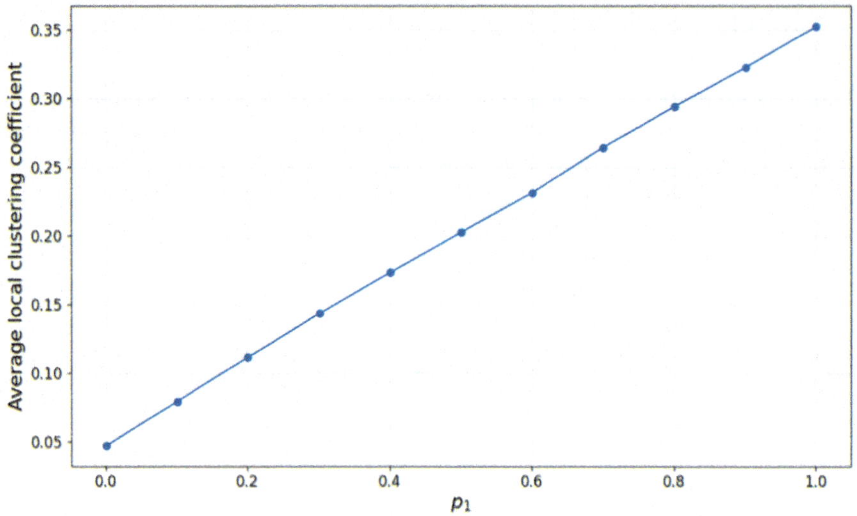

Fig. 1 Controlling $\overline{C}$ by p_1: there is a linear relation between both measures ($m = 5$, $N = 50000$)

Table 2 Generated networks and their measures for different values of m ($p_1 = 0.5$, $N = 20000$). Parameters: number of edges added at each iteration m. Measures: average local clustering coefficient $\overline{C}$; transitivity C; average shortest path length $\overline{L}$; estimated power-law exponent γ; maximum degree of the network d_{max}

m	$\overline{C}$	C	$\overline{L}$	γ	d_{max}
1	0.1358	0.0795	6.5044	−2.5546	121
2	0.3628	0.0951	5.3560	−2.3339	202
3	0.3239	0.0951	4.7796	−2.0969	232
4	0.2622	0.0847	4.3629	−2.2439	282
5	0.2125	0.0718	4.0473	−2.0709	310
6	0.1717	0.0635	3.8421	−2.1094	348
7	0.1467	0.0559	3.6765	−2.1879	408
8	0.1265	0.0508	3.5599	−2.0339	444
9	0.1123	0.0459	3.4487	−2.0735	494
10	0.1007	0.0416	3.3456	−2.0741	505

Table 2 shows that both transitivity and the average shortest path length decrease as m increases, while γ remains relatively stable, fluctuating between −2.00 and −2.30 when $m \geq 2$. Figure 2 demonstrates this quick decay for the average shortest path length.

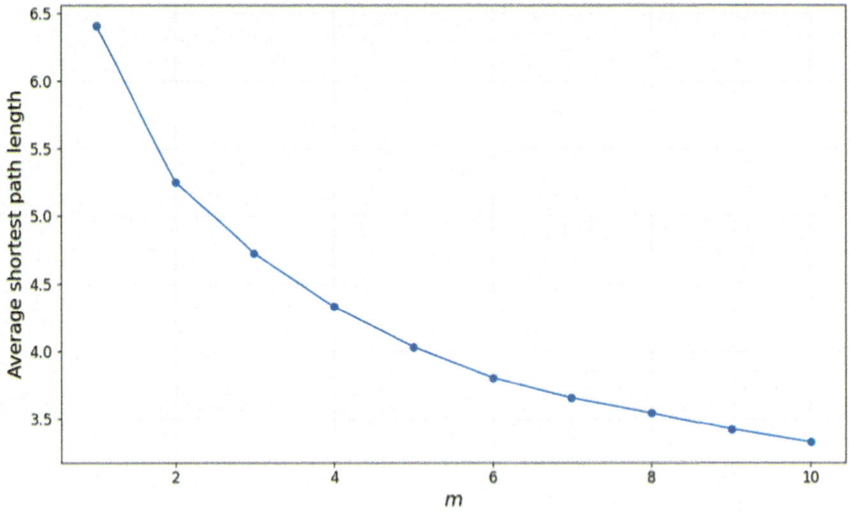

Fig. 2 Controlling $\overline{L}$ by m: the average distance quickly drops down as m increases ($p_1 = 0.5$, $N = 20000$)

Table 3 Generated networks and their measures for different values of N ($m = 5$, $p_1 = 0.5$). Parameter: total number of nodes N added to the graph. Measures: average local clustering coefficient $\overline{C}$; transitivity C; average shortest path length $\overline{L}$; estimated power-law exponent γ; maximum degree of the network d_{max}

N	$\overline{C}$	C	$\overline{L}$	γ	d_{max}
100	0.2949	0.2116	2.3636	−0.9623	35
200	0.2725	0.1668	2.5443	−1.0640	48
1000	0.2327	0.1144	3.1492	−1.8068	89
2000	0.2226	0.0974	3.3488	−1.8760	136
5000	0.2100	0.0675	4.3622	−2.2204	433
10000	0.2515	0.0807	4.4231	−2.2239	594
20000	0.2577	0.0852	4.4957	−2.2362	484
50000	0.2460	0.0842	4.6178	−2.3176	559
100000	0.2340	0.0796	4.7349	−2.3260	678

Finally, Table 3 shows the characteristics of the generated networks for different values of N, the total number of added nodes, fixing $m = 5$ and $p_1 = 0.5$. It can be seen that after $N = 10.000$, the clustering coefficients vary little, remaining at a significantly high level. Table 3 also shows the relationship between γ, and N, where the first decreases as the second increases.

Moreover, Fig. 3 and Table 3 illustrate that the average shortest path length $\overline{L}$ grows proportionally to the logarithm of N (small-world behavior), proving that the algorithm can create networks with short distances between the nodes.

For comparison, Table 4 shows the average local clustering coefficients and the

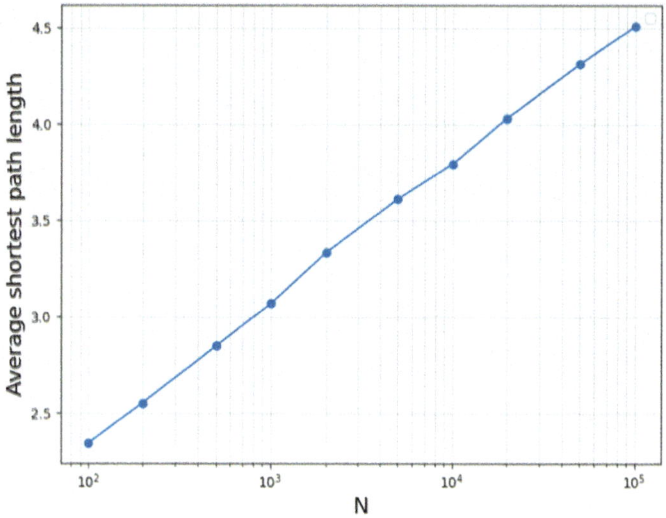

Fig. 3 Relationship between the number of nodes N, plotted in logarithmic scale, and the average shortest path length $\overline{L}$ ($m = 5$, $p_1 = 0.5$)

Table 4 Comparison of networks generated by the Herrera and Zufiria model (1) and our model (2). Measures: average local clustering coefficient $\overline{C}$; average shortest path length $\overline{L}$

N	m	$\overline{C_1}$	$\overline{C_2}$	$\overline{L_1}$	$\overline{L_2}$
20000	2	0.4248	0.3734	10.2354	5.2538
20000	3	0.4020	0.3286	7.3107	4.7691
20000	4	0.3126	0.2505	6.2514	4.3309
50000	2	0.4218	0.3694	11.4507	5.6787
50000	3	0.3960	0.3271	9.1531	5.0849
50000	4	0.3130	0.2643	6.1754	4.6525
70000	2	0.4200	0.3676	11.9726	5.8315
70000	3	0.3963	0.3254	8.5344	5.1880
70000	4	0.3140	0.2621	6.0804	4.7553

average distances for networks generated by our algorithm and those generated without adding the special edge, equivalently to Herrera and Zufiria's model in which the nodes have a probability p_1 of making $l = 1$. The distances in our model are significantly lower, especially for smaller values of m, following the well-known *six degrees of separation* principle.

5 Conclusions

The study of complex networks gained prominence across various scientific disciplines [15–17]. In this study, we presented a simple and flexible algorithm that generates a wide range of networks with different values of several network measures, such as clustering coefficients, average shortest path length, and the estimated power-law coefficient, without employing global information about the graph topology or degree distributions.

The proposed approach can generate networks with long-tailed degree distributions, high clustering coefficients, and short average distances between the nodes, three of the fundamental characteristics of real-world networks. Using this simple random-walk-based approach, it is possible to generate networks with structural characteristics similar to those found in real-world networks without using hyperbolic geometry methods [18]. The implementation of our algorithm is publicly available on a GitHub repository [19].

We can control the average distances between network nodes, keeping them small and logarithmic in the size of the network, a neglected aspect in previous models [10, 14]. In addition, our model allows different probability distributions to set up the distance value d used to introduce a degree of randomness in the network connections. By choosing the appropriate distribution, the average shortest path length $\overline{L}$ can be made smaller or larger, even for large graphs with tens and hundreds of thousands of nodes.

Network clustering coefficients grow proportionally to the parameter p_1, reaching fairly high values while keeping realistic the other measures: average distances and "long-tailness".

In future works, we intend to explore different distributions to control the value of l, the number of steps between any two nodes marked successively. We will study the possibility of increasing the values of l, exploring the impact of the chosen distribution on the clustering coefficients and the average distances between nodes.

We are also interested in incorporating a fourth characteristic of real-world networks into our model: a high degree of modularity. High modularity implies the presence of modules (groups, communities) of nodes, with more dense connections within modules but sparse connections between nodes in different modules. Incorporating this feature organically, without being forced or planned in advance, seems like an interesting challenge to tackle.

Acknowledgements Ruben Interian was supported by research grant PIND 2423/24 (Universidade Estadual de Campinas). This study was financed, in part, by the São Paulo Research Foundation (FAPESP), Brasil. Process Number 2024/12936-5.

References

1. de Arruda, G.F., Rodrigues, F.A., Moreno, Y.: Fundamentals of spreading processes in single and multilayer complex networks. Phys. Rep. **756**, 1–59 (2018)
2. Interian, R., Marzo, R.G., Mendoza, I., Ribeiro, C.C.: Network polarization, filter bubbles, and echo chambers: an annotated review of measures and reduction methods. Int. Trans. Oper. Res. **30**(6), 3122–3158 (2023)
3. Seguin, C., Sporns, O., Zalesky, A.: Brain network communication: concepts, models and applications. Nat. Rev. Neurosci. **24**(9), 557–574 (2023)
4. Zhou, B., Holme, P., Gong, Z., Zhan, C., Huang, Y., Lu, X., Meng, X.: The nature and nurture of network evolution. Nat. Commun. **14**(1), 7031 (2023)
5. Interian, R.: A political radicalization framework based on Moral Foundations Theory. Mathematics **12**(13) (2024)
6. Barabási, A.-L., Albert, R.: Emergence of scaling in random networks. Science **286**(5439), 509–512 (1999)
7. Broido, A., Clauset, A.: Scale-free networks are rare. Nat. Commun. **10**(1), 1017 (2019)
8. Bhattacharya, S., Sinha, S., Roy, S.: Impact of structural properties on network structure for online social networks. In: Procedia Computer Science. International Conference on Computational Intelligence and Data Science, vol. 167, pp. 1200–1209 (2020)
9. Watts, D., Strogatz, S.: Collective dynamics of 'small-world' networks. Nature **393**(6684), 440–442 (1998)
10. Saramäki, J., Kaski, K.: Scale-free networks generated by random walkers. Physica A: Stat. Mech. Its Appl. **341**, 80–86 (2004)
11. Boguñá, M., Bonamassa, I., de Domenico, M., Havlin, S., Krioukov, D., Serrano, M.A.: Network geometry. Nat. Rev. Phys. **3**(2), 114–135 (2021)
12. Boguñá, M., Krioukov, D., Almagro, P., Serrano, M.A.: Small worlds and clustering in spatial networks. Phys. Rev. Res. **2**, 023040 (2020)
13. Holme, P., Kim, B.J.: Growing scale-free networks with tunable clustering. Phys. Rev. E **65**(2) (2002)
14. Herrera, C., Zufiria, P.J.: Generating scale-free networks with adjustable clustering coefficient via random walks. In: 2011 IEEE Network Science Workshop, pp. 167–172 (2011)
15. Boccaletti, S., De Lellis, P., del Genio, C.I., Alfaro-Bittner, K., Criado, R., Jalan, S., Romance, M.: The structure and dynamics of networks with higher order interactions. Phys. Rep. **1018**, 1–64 (2023)
16. Interian, R., Rodrigues, F.A.: Group polarization, influence, and domination in online interaction networks: a case study of the 2022 Brazilian elections. J. Phys. Complex. **4**(3), 035008 (2023)
17. Scabini, L., Ribas, L.C., Neiva, M.B., Junior, A., Farfán, A., Bruno, O.M.: Social interaction layers in complex networks for the dynamical epidemic modeling of COVID-19 in Brazil. Physica A: Stat. Mech. Its Appl. **564**, 125498 (2021)
18. Krioukov, D., Papadopoulos, F., Kitsak, M., Vahdat, A., Boguñá, M.: Hyperbolic geometry of complex networks. Phys. Rev. E **82**, 036106 (2010)
19. Morais, J.P.C., Interian, R.: Github repository: real-world networks algorithm (2024). https://github.com/jpcmorais16/real-world-network-algorithm

A Framework for Efficient Estimation of Closeness Centrality and Eccentricity in Large Networks

Patrick C. Trindade, Maximilien Dreveton, and Daniel R. Figueiredo

Abstract Centrality indices, such as closeness and eccentricity, are key to identifying influential nodes within a network, with applications ranging from social and biological networks to communication and transportation systems. However, computing these indices for every node in large graphs is computationally prohibitive due to the need for solving the All-Pairs Shortest Path (APSP) problem. This paper introduces a framework for approximating closeness and eccentricity centrality by selecting a sequence of strategically chosen anchor nodes, from which Breadth-First Searches (BFS) are performed. We present two anchor-selection strategies that minimize estimation error for these indices and evaluate their effectiveness on synthetic and real-world networks. Comparative results indicate that while random anchor selection occasionally achieves lower errors for closeness, other strategies outperform in eccentricity estimation. This study highlights the effectiveness of anchor-based approximations and the trade-offs between different selection methods in estimating centrality at scale.

Keywords Network centrality · Closeness · Eccentricity · Approximation algorithm

P. C. Trindade (✉) · D. R. Figueiredo
Systems Engineering and Computer Science (PESC), Federal University of Rio de Janeiro (UFRJ), Rio de Janeiro, Brazil
e-mail: patricktrindade@poli.ufrj.br

D. R. Figueiredo
e-mail: daniel@cos.ufrj.br

M. Dreveton
School of Computer and Communication Sciences (IC), EPFL, Lausanne, Switzerland
e-mail: maximilien.dreveton@epfl.ch

M. Macedo et al. (eds.), *Complex Networks XVI*, Springer Proceedings in Complexity,
https://doi.org/10.1007/978-3-031-93619-7_2

1 Introduction

Network centrality indices, such as closeness centrality and eccentricity, are crucial in graph theory and network analysis because they offer insights into the relative importance, influence, or accessibility of nodes within a network [1, 3]. By quantifying aspects like a node's proximity to others (closeness centrality) or its maximum distance to any other node (eccentricity), these indices help identify key vertices that might act as hubs, bridges, or bottlenecks within the network. In social networks, for instance, nodes with high closeness centrality are likely to spread information more efficiently, while those with low eccentricity are part of the network's core set of nodes. Similarly, in transportation, biological, or communication networks, understanding which nodes hold central positions can inform strategies for optimizing connectivity, controlling information flow, or fortifying against disruptions, respectively.

For a connected graph $G = (V, E)$ with $n = |V|$ vertices, the *closeness centrality* of a node $u \in V$ is defined by

$$\mathrm{c}_u = \frac{n-1}{\sum_{v \in V} d_u(v)}$$

where $d_u(v)$ is the shortest-path distance between u and v in G. The *eccentricity* of $u \in V$ is defined by

$$\mathrm{e}_u = \max_{v \in V} d_u(v).$$

In many scenarios, network nodes are to be ranked (i.e., sorted) according to some network centrality indices. This allows us to determine the top-ranked or low-ranked nodes according to some index, for example. In order to obtain a full ranking (i.e., ordering) of the nodes, the centrality index must be computed for all network nodes. However, computing these indices for each node in a large graph is computationally intensive, as it involves finding the shortest paths between all pairs of vertices. This problem, known as the All-Pairs Shortest Path (APSP) problem, is a classic challenge in computer science. Running a Breadth First Search (BFS) from each node solves the APSP with running time complexity of $\Theta(nm)$, where $m = |E|$ is the number of edges in graph [6].

Solving APSP requires significant computational resources as many real networks continue to grow in size like social networks, transportation networks, or information networks. To address this scalability challenge, researchers have developed approximate algorithms, parallel algorithms, and heuristic-based algorithms to compute different centrality indices more efficiently, circumventing the need to solve the classic APSP problem. In what follows, a brief discussion of the proposed algorithms is presented.

Related works. Cohen et al. [4] propose an approximation algorithm based on sampling and probabilistic techniques, specifically leveraging adaptive sampling methods to estimate closeness centrality efficiently without requiring the full computation of all-pairs shortest paths. The algorithm significantly reduces computational complexity by focusing on a subset of vertices and incrementally building estimates that converge to a close approximation of true closeness centrality. The paper also includes detailed theoretical guarantees on the accuracy of these estimates.

Bergamini et al. [2] presents an efficient algorithm for finding the top-k nodes with the highest closeness centrality in unweighted graphs, rather than calculating closeness centrality for all nodes. Their approach combines a pruning strategy with a bidirectional breadth-first search (BFS) to avoid unnecessary calculations. By leveraging upper and lower bounds on closeness centrality scores as they explore the graph, the algorithm can discard certain nodes early from consideration, thus improving efficiency. The method also integrates efficient data structures (such as min-heap) to track and update centrality scores, which further speeds up the identification of top-k nodes.

Our approach. This work presents a framework for estimating the closeness and eccentricity of nodes. The main idea is to dynamically determine a sequence of nodes (known as anchors) on which a BFS will be executed. Using the results of these BFS, the closeness and eccentricity of all nodes can be estimated. The goal is to determine a sequence of anchors that has a low error with respect to the true values for closeness and eccentricity. This work proposes two approaches based on different intuitions to determine the sequence of anchors (discussed in detail in Sect. 2). These approaches were implemented and executed in two real networks and two network models, comparing the results with random strategy and the approach by Cohen et al. [4]. Interestingly, results indicate that random often exhibits a lower error than the alternatives for closeness estimation but not for eccentricity. This finding highlights the lack of a clear approach that is consistently superior in estimating both closeness and eccentricity.

Structure of the paper. The approach and estimators for centrality and eccentricity proposed in this work are presented in Sect. 2. Section 3 presents the numerical evaluation and a comparison of different approaches. Finally, a brief conclusion for the paper is presented in Sect. 4.

2 Centrality Estimators

Recall that closeness and eccentricity for a node $u \in V$ can be determined by running a Breadth First Search (BFS) on u since this provides the distance $d_u(v)$ between node u and every other node $v \in V$. However, this BFS also provides information about other nodes, since $d_v(u) = d_u(v)$ (the graph under consideration is undirected). The key idea behind the proposed estimators is to consider a sequence of nodes to run a BFS, called anchor nodes. The closeness and eccentricity for anchor nodes are precisely determined, while the closeness and eccentricity for non-anchor nodes are estimated.

Let $A_t = (a_1, \ldots, a_t)$ denote a sequence of t anchor nodes. Assume that node v is not an anchor. Note that each anchor node a_i, $1 \leq i \leq t$ provides a distance to v, namely $d_{a_i}(v)$. Thus, an estimator for the closeness of node v can be designed as follows:

$$\hat{c}_v(t) = \frac{t}{\sum_{a \in A_t} d_a(v)} \tag{1}$$

As the anchor set size increases, $\hat{c}_v(t)$ will intuitively approximate c_v and equality will always be established when $A_t = V \setminus \{v\}$.

Similarly, an estimator for the eccentricity of node v can be designed as follows:

$$\hat{e}_v(t) = \max_{a \in A_t} d_a(v) \tag{2}$$

Again, as the anchor set size increases, $\hat{e}_v(t)$ will intuitively approximate e_v and equality will always be established when $A_t = V \setminus \{v\}$.

Note that the quality of the estimators $\hat{c}_v$ and $\hat{e}_v$ will intuitively depend on the sequence of anchors A_t. In particular, these estimators can have a bias and errors that depend on the anchors and the network. Thus, a fundamental question concerns how to efficiently choose a sequence of anchors to yield more accurate estimators. Moreover, the sequence of anchors can be determined in an online fashion: the choice of anchor a_{t+1} can depend on the distances obtained by all previously chosen anchors $a_1, \ldots, a_t$.

We propose three strategies to determine the choice of anchors: a simple random selection and two online strategies.

2.1 *Largest Average Distance*

The *Largest Average Distance* (LAD) strategy aims to cover the periphery of the network by choosing the next anchor to be the node with the largest average distance to existing anchors. Note that a higher average distance indicates weaker closeness centrality. LAD proceeds as follows:

1. **First anchor**: Randomly select the first anchor node.
2. **Subsequent anchors**: For each remaining non-anchor node v, calculate $\hat{c}_v(t)$ and choose the node with the largest average distance as the next anchor. In particular,

$$a_{t+1} = \arg \max_{v \in V \setminus A_t} \hat{c}_v(t) \tag{3}$$

If more than one node attains the maximum value, then a random node is chosen among them.

2.2 Largest Minimum Distance

The *Largest Minimum Distance* (LMD) strategy aims to spread the anchors evenly across the network, providing a more balanced distribution of anchor nodes. LMD follows these steps:

1. **Initial anchor**: Randomly select the first anchor.
2. **Subsequent Anchors**: For each remaining non-anchor node v, calculate the minimum distance to all anchor nodes, namely

$$\delta_v(t) = \min_{a \in A_t} d_a(v). \tag{4}$$

   Choose as the next anchor the node with the largest $\delta_v(t)$. In particular,

$$a_{t+1} = \arg \max_{v \in V \setminus A_t} \delta_v(t) \tag{5}$$

   If more than one node attains the maximum value, then a random node is chosen among them.

Note that both LAD and LMD are a randomized algorithms as they use randomness on their first step and to break possible ties when selecting anchor nodes. Thus, independent executions of LAD and LMD on the same network is likely to produce different sequences of anchors, respectively.

2.3 Random

The *Random* (RND) approach provides a baseline by selecting the sequence of anchors uniformly at random over. Thus, a_{t+1} is a node chosen uniformly at random from the set $V \setminus A_t$.

Figure 1 provides an illustration of the set of anchors selected by the different estimators. Note that LMD will select as anchor a node that is far from the current set of anchors. This will intuitively spread the anchors across the network, including

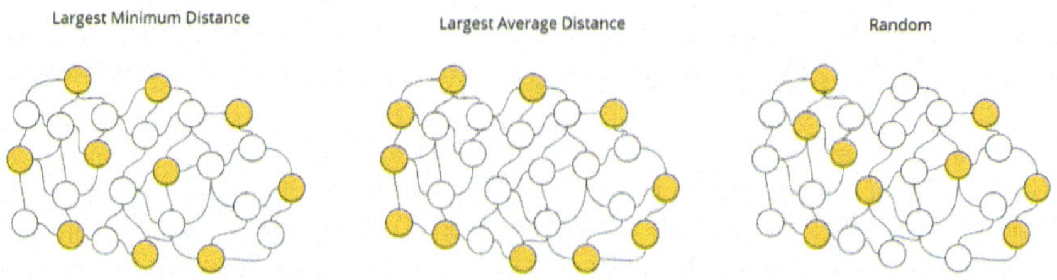

Fig. 1 Set of anchors selected according to the different estimators

its core. In contrast, LAD will select as anchor nodes that have a large average distance to other anchors, thus avoiding the core of the network (where the average distance to anchors is relatively small). Intuitively, LAD selects anchors on the periphery of the network. Last, under the random approach every node is equally likely to be chosen. Thus, the likelihood of choosing a node from the core or the periphery depends on the ratio between the number core and peripheral nodes.

2.4 Computational Complexity

Recall that running a single BFS on a connected graph G has run time complexity $\Theta(m)$ where $m = |E|$ is the number of edges of G (since m is always greater than $n = |V|$ when G is connected). Thus, in order for the estimators to be computationally efficient, the time required to choose an anchor should be small.

The RND approach requires constant time $\theta(1)$ to make a random selection on the set of non-anchor nodes. Thus, the time required to determine t anchors and run the corresponding BFS is $\Theta(tm)$

However, both LAD and LMD require time $\theta(|V \setminus A_t|)$ to determine the anchor a_{t+1} which in the worst case (i.e., selecting the second anchor) requires time $\Theta(n)$ since the maximum value must be determined among the non anchor nodes. Thus, the running time to determine the anchors and run the corresponding BFS is $\Theta(t(n + m)) = \Theta(tm)$ since $n = O(m)$ assuming G is connected.

While the theoretical worst case complexity of the strategies for selecting the anchors and running the corresponding BFS is identical, in practice the running time is different (to be shown later). Moreover, there seems to be an inherent tradeoff between running more BFS or better selecting the next anchor node (and running fewer BFS) that is not addressed in this work.

3 Numerical Experiments

3.1 Datasets and Performance Metrics

The numerical evaluation was carried out using both synthetic and real networks. For synthetic networks, the Erdös-Rènyi (ER) [1] and Artificial Benchmark for Community Detection (ABCD) [5] random graph models were used. In the ER model, the number of nodes was $n = 4000$ and edge probability $p = 0.01$. For the ABCD model, a graph with $n = 4000$ nodes were generated and node degree followed a power-law distribution with an exponent of 2.5, and the graph is divided into $k = 20$ communities with sizes also following a power-law distribution with an exponent of 2.0. For real-world data, two graphs from the SNAP repository [8] were considered:

Table 1 Network statistics for the networks under investigation

Dataset	$n = \|V\|$	$m = \|E\|$	$\bar{b}$	Triangles	Diameter
ER	4000	79,564	39.78	31,524	4
ABCD	4000	82,698	41.35	79,929	4
Facebook	4039	88,234	43.7	4,836,030	8
Arxiv GR-QC	4158	13,422	6.5	143,337	17

- the Arxiv GR-QC network [7], a scientific collaboration network from ArXiv papers with co-authors who have submitted papers to the General Relativity and Quantum Cosmology category;
- a friendship network among Facebook users [9].

Table 1 provides a summary of key network statistics for each network.

We report the error with the mean squares error and the Jaccard similarity of the top 1% of the nodes. More precisely, let c and $\hat{c}(t)$ be the true and estimated closeness centrality once t anchors have been considered, respectively. The mean-squared error (MSE) is defined as

$$\text{MSE}(c, \hat{c}(t)) = \frac{1}{n} \sum_{u \in V} \left(c_u - \hat{c}_u(t)\right)^2 = \frac{1}{n} \sum_{u \in V \setminus A_t} \left(c_u - \hat{c}_u(t)\right)^2$$

Note that once a node u is chosen to be an anchor node, then $\hat{c}_u(t) = c_u$, thus only non-anchor nodes contribute to the MSE.

Recall that the Jaccard index between two sets A and B is defined as $\text{J}(A, B) = \frac{|A \cap B|}{|A \cup B|}$. We define the top 1% elements of a vector x with n entries, we first sort the elements of x in descending order. The top 1% elements are then the largest $\lceil 0.01 \cdot n \rceil$ elements of this sorted vector. We denote this set of top 1% elements as $\text{top1}(x)$. The top-1% Jaccard similarity between c and $\hat{c}(t)$ is defined as

$$\mathbf{J}_{\text{top1}}(c, \hat{c}(t)) = \text{J}\left(\text{top1}(c), \text{top1}(\hat{c}(t))\right).$$

We define analogous quantities $\text{MSE}(e, \hat{e}(t))$ and $\mathbf{J}_{\text{top1}}(e, \hat{e}(t))$ for eccentricity.

3.2 Closeness

To evaluate the closeness centrality estimators, we compared our anchor-based methods (LAD, LMD, and RND) against the approximation method currently implemented in the NetworKit library[1] [10]. This provides a benchmark for assessing

[1] See https://networkit.github.io/dev-docs/python_api/centrality.html. The algorithm implemented is proposed in [4].

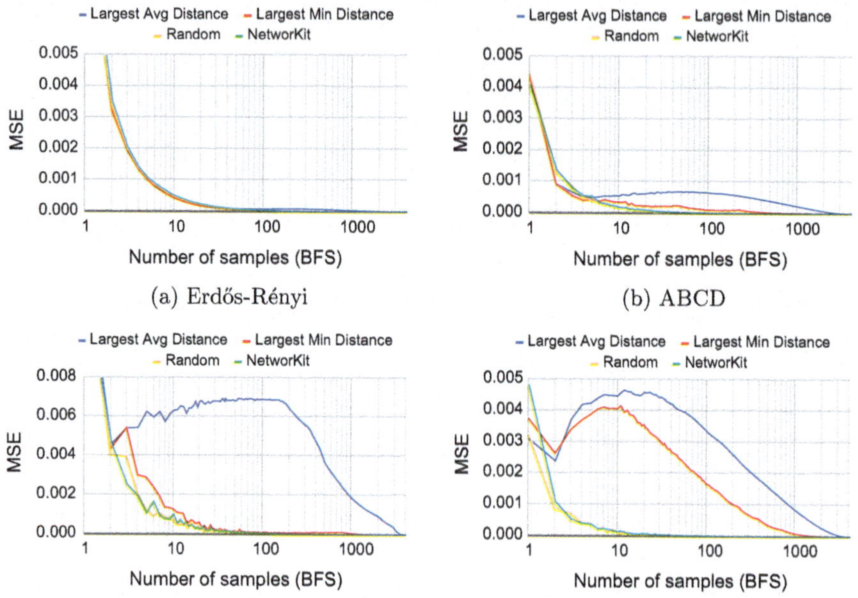

(a) Erdős-Rényi (b) ABCD

(c) Facebook (d) Arxiv GR-QC

Fig. 2 Mean square error of the estimated closeness centrality by the different estimators on four data sets

the accuracy and efficiency of our methods relative to an established approximation approach.

In each experiment, we run all four methods on synthetic and real-world networks, measuring performance through mean-squared error (MSE) and Jaccard similarity for the top 1% of nodes by closeness centrality. Results are given in Figs. 2 and 3.

Synthetic graphs. In synthetic networks generated by the Erdös-Rényi and ABCD models, all methods demonstrated very similar MSE and Jaccard similarity. Moreover, the performance of Erdös-Rényi and ABCD are also very similar. This suggests that the structure of these synthetic networks does not strongly favor any particular anchor selection strategy. We believe this behavior occurred because these synthetic graphs lack nodes that are very far from the core of the graph, and thus minimizing the performance differences among strategies.

Real-world networks. In contrast, notable differences appeared in the two real-world networks. The NetworKit and RND methods outperformed LAD and LMD, yielding lower MSE and higher Jaccard similarity. This pattern suggests that NetworKit and RND provided more balanced network coverage, while LAD and LMD introduced biases by selecting anchors with high average or minimum distances, often situated in peripheral communities. Such communities are typically less connected to the network's core, resulting in inflated closeness estimates in our approximation. Interestingly, the MSE decays faster for RND and NetworKit on the Arxiv GR-QC

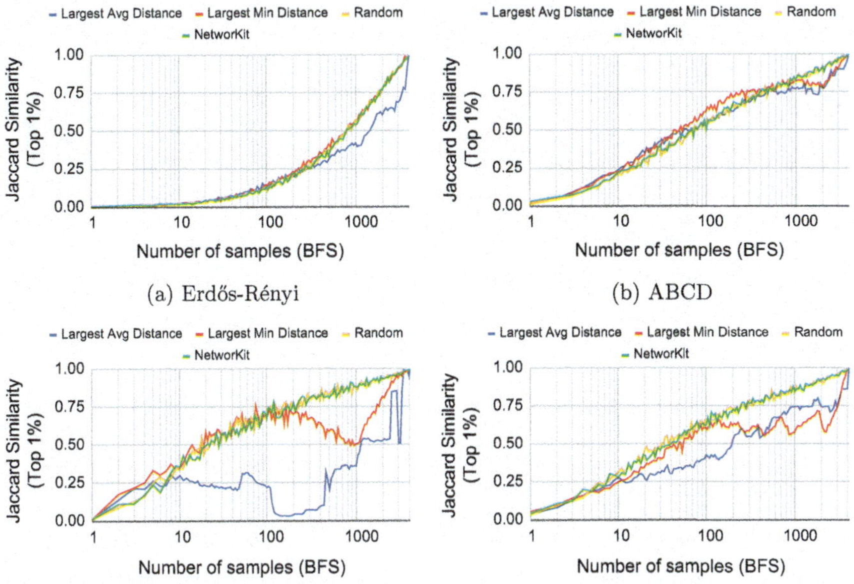

(a) Erdős-Rényi (b) ABCD (c) Facebook (d) Arxiv GR-QC

Fig. 3 Jaccard similarity of the estimated closeness centrality of the top 1% vertices obtained by the different estimators on four data sets

network than in any other network. In particular, with just 3 anchors the MSE is below 0.001 in this case. Moreover, the Jaccard similarity for RND and NetworKit strategies grows much faster on real networks than on synthetic networks. This indicates that finding the top 1% of nodes using an approximate method is much more effective on real networks, again possibly due to their diversity in degree and network structure. In particular, with just 50 anchors the Jaccard similarity is above 0.5 for both real networks when using RND or NetworKit.

Summary. The tendency of LAD and LMD to select distant nodes reflects their design to maximize average or minimum distance, respectively. However, this bias can cause overestimation of closeness centrality in real-world networks. The unbiased RND method and NetworKit approach deliver more reliable estimates in these cases. This suggests that while LAD and LMD can effectively maximize spatial diversity, they may not be optimal for closeness centrality approximation in networks with pronounced community clustering, where balanced network coverage is essential.

Finally, we provide in Fig. 4 the execution time of all algorithms. While RND typically exhibits slower empirical running times, it achieves MSE and Jaccard similarity results comparable to those of the NetworKit implementation.

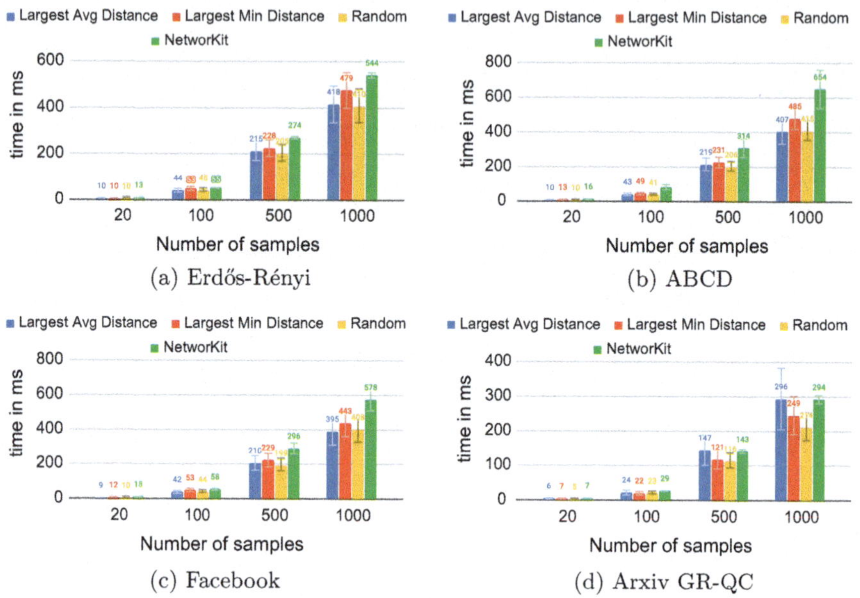

(a) Erdős-Rényi (b) ABCD (c) Facebook (d) Arxiv GR-QC

Fig. 4 Execution time of the different closeness centrality estimators on four data sets

3.3 *Eccentricity*

For the eccentricity centrality estimators, we measured each method's performance on both synthetic and real-world graphs using MSE and Jaccard similarity for the top 1% of nodes with the highest eccentricity. Results are given in Figs. 5 and 6.

Synthetic graphs. For synthetic graphs, the three strategies have very similar performance for the first 100 anchors or so. After 100 anchors, the LAD strategy performs better than LMD and RND. Interestingly, the MSE for all estimators decreases faster on Erdös-Rényi graphs than on ABCD graphs, but with both exhibiting a long tail until the MSE reaches zero. This slower convergence suggests that the lack of clear peripheral nodes in these graphs makes it more challenging for anchor-based methods to rapidly identify nodes with the largest eccentricity. Indeed, the Jaccard similarity with the top 1% is zero for up to 100 anchors or so. In this metric, it is clear that LAD has an advantage over the other strategies.

Real-world networks. On the real-world networks, where network structure is far from uniform, nodes with large eccentricity typically stand out and can be more easily identified. Indeed, the LAD and LMD strategies achieved perfect Jaccard similarity (a value of 1) with around 10 anchor nodes or so. This rapid convergence indicates that selecting nodes with large distances from previous anchors efficiently covers the network extremities, making it easier to identify nodes with large eccentricity. Note

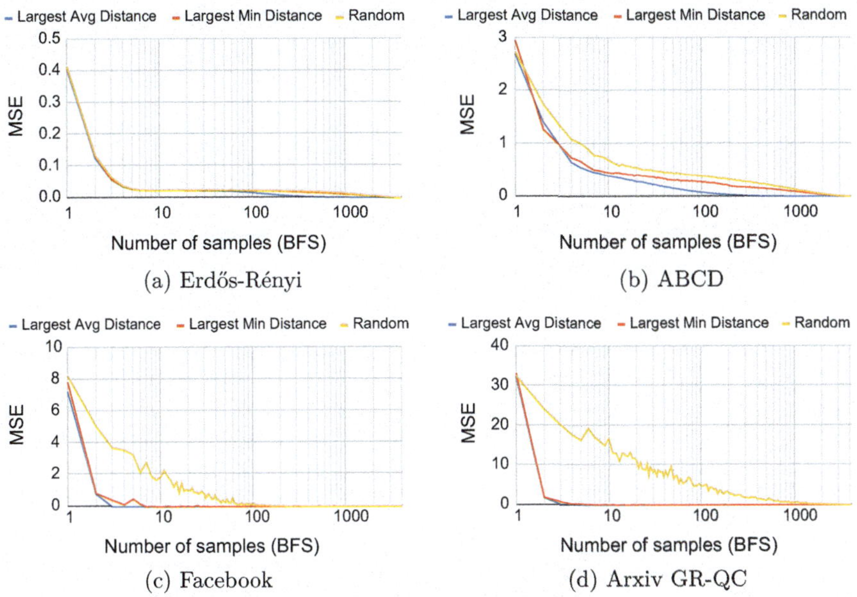

(a) Erdős-Rényi (b) ABCD (c) Facebook (d) Arxiv GR-QC

Fig. 5 Mean square error of the estimated eccentricity by the different estimators on four data sets

that RND converges much more slowly, requiring more samples to approach similar levels of Jaccard similarity.

Summary. While LAD and LMD are less effective than RND for closeness estimation due to their bias toward peripheral nodes, these methods offer significant advantages for eccentricity estimation, in particular LAD. By focusing on distant nodes, they effectively capture the extremities of the network, which is particularly useful in real-world graphs where identifying the outermost nodes is crucial for accurately determining nodes with large eccentricity. Overall, the experiments show that LAD is an efficient and reliable eccentricity estimates.

4 Conclusion

Efficient calculation of centrality metrics becomes a fundamental problem as networks continue to grow in size. In particular, classic and exact algorithms for computing closeness and eccentricity for all network nodes become unfeasible (in running time) on networks with billions of nodes (i.e., exact solution of the All-Pairs Shortest Path problem). An approach to tackle this problem the design of approximate algorithms to estimate such centrality metrics.

This work proposes a framework for approximating closeness and eccentricity centrality metrics using an anchor-based approach that accommodates different

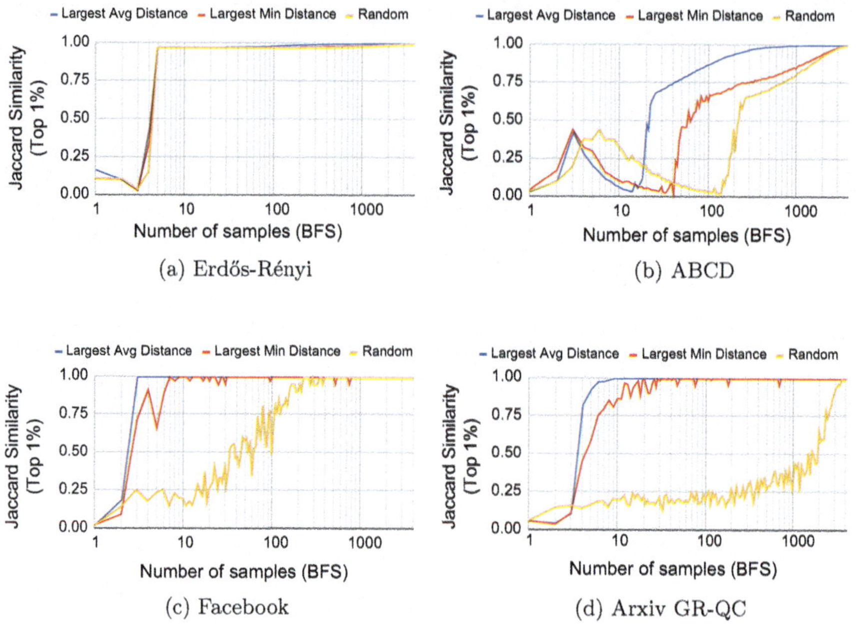

(a) Erdős-Rényi (b) ABCD

(c) Facebook (d) Arxiv GR-QC

Fig. 6 Jaccard similarity of the estimated eccentricity of the top 1% vertices obtained by the different estimators on four data sets

anchor selection strategies. Once an anchor is selected, a Breadth-First Searches (BFS) is executed on the anchor and computes distances from the anchor to all others. By iteratively selecting the anchors using the distance information obtained by prior anchors, closeness and eccentricity can be estimated for all nodes. This work considers two strategies (LAD and LMD) and the random strategy as baseline.

Experimental results considering the MSE and Jaccard similarity with the top 1% across both synthetic and real-world networks demonstrated the effectiveness of the proposed approach. Interestingly, results considering synthetic networks show little difference with respect to the different strategies. However, when considering real networks, different strategies yield very different results. For estimating closeness, the random strategy showed competitive performance to a state-of-the-art method and outperformed LAD and LMD. However, LAD and LMD showed superior performance (both MSE and Jaccard similarity) for estimating eccentricity (with an edge for LAD).

These findings underscore the potential of strategic anchor selection to improve the accuracy of centrality approximations in large-scale networks. Future work may explore optimizing anchor strategies further or applying this framework to other centrality indices to broaden its applicability across diverse network types.

References

1. Avrachenkov, K., Dreveton, M.: Statistical Analysis of Networks. Boston-Delft: now publishers (2022)
2. Bergamini, E., Borassi, M., Crescenzi, P., Marino, A., Meyerhenke, H.: Computing top-k closeness centrality faster in unweighted graphs. Trans. Knowl. Discov. Data (TKDD) **13**(5), 1–40 (2019)
3. Boldi, P., Vigna, S.: Axioms for centrality. Internet Math. **10**(3–4), 222–262 (2014)
4. Cohen, E., Delling, D., Pajor, T., Werneck, R.F.: Computing classic closeness centrality, at scale. In: Conference on Online Social Networks, pp. 37–50 (2014)
5. Kamiński, B., Prałat, P., Théberge, F.: Artificial benchmark for community detection (ABCD)-fast random graph model with community structure. Netw. Sci. **9**(2), 153–178 (2021)
6. Kleinberg, J., Tardos, E.: Algorithm Design. Pearson Education (2006)
7. Leskovec, J., Kleinberg, J., Faloutsos, C.: Graph evolution: densification and shrinking diameters. Trans. Knowl. Discov. Data **1**(1) (2007)
8. Leskovec, J., Krevl, A.: SNAP datasets: Stanford large network dataset collection (2014). http://snap.stanford.edu/data
9. Leskovec, J., Mcauley, J.: Learning to discover social circles in ego networks. Adv. Neural Inform. Process. Syst. **25** (2012)
10. Staudt, C., Sazonovs, A., Meyerhenke, H.: NetworKit: an interactive tool suite for high-performance network analysis, p. 14 (2014). CoRR arXiv:1403.3005

Network-Centric Analysis of Memetic Operators and Communication Topologies for Swarm Intelligence Algorithms

Carlos Silva, Clodomir Santana, Ronaldo Menezes, and Carmelo Bastos-Filho

Abstract Population-based optimisers, such as swarm intelligence and evolutionary algorithms, have been widely applied across various optimisation problems due to their flexibility, efficiency, and balance between exploration and exploitation. However, they can be prone to premature stagnation and suboptimal convergence, especially in complex search spaces. Memetic algorithms that combine Particle Swarm Optimization (PSO) with local search operators have been introduced to address these issues. This paper presents a network-based comparative study of memetic operators for swarm optimisers. As a case study, we selected three memetic PSO variants: PSO with Pattern Search, PSO with Hill Climbing, and PSO with Simulated Annealing. We also assessed three communication topologies for PSO (e.g., Global, Local, and Von Neumann). Using Interaction Networks and metrics such as Portrait Divergence and Interaction Diversity, we model and assess these variants' convergence behaviour, and exploration-exploitation dynamics over time. Results indicate that the influence of the memetic operators and communication topologies affects different aspects of the network, such as connection patterns, the presence of hubs and clusters, and the edges' weights. Additionally, the network analysis offers valuable insights into the exploration-exploitation balance, convergence speed, and the role of topological structures in shaping swarm dynamics.

Keywords Complex networks · Metaheuristics · Particle swarm optimisation · Interaction networks · Memetic algorithms

C. Silva (✉) · C. Bastos-Filho
University of Pernambuco, Recife, Brazil
e-mail: cabs@ccomp.poli.br

C. Bastos-Filho
e-mail: carmelofilho@ieee.org

C. Santana
University of California, Davis, USA
e-mail: clsantana@ucdavis.edu

R. Menezes
University of Exeter, Exeter, UK
e-mail: r.menezes@exeter.ac.uk

M. Macedo et al. (eds.), *Complex Networks XVI*, Springer Proceedings in Complexity,
https://doi.org/10.1007/978-3-031-93619-7_3

1 Introduction

Population-based metaheuristics, such as Swarm Intelligence (SI) algorithms, find applications in various fields like engineering, data science, and economics [17]. These algorithms utilise individuals or particles that navigate the search space, exchanging information and adapting their positions based on personal and collective knowledge. In these algorithms, interaction is crucial for practical exploration and exploitation of the search space [21].

However, traditional SI algorithms can struggle with complex or high-dimensional search spaces, leading to slow convergence and prolonged runtimes [21]. Additionally, they may become trapped in local optima, hindering the exploration of potentially better solutions. To tackle these issues, researchers have developed adaptations and enhancements, including new operators and mechanisms [15].

Among the solutions proposed to enhance SI, integrating local search operators has proven effective [3]. These operators refine candidate solutions by conducting focused searches around promising regions, mitigating premature convergence and improving solution quality [16]. While the impact of individual operators has been studied, the interplay of multiple operators and their collective influence on swarm-based algorithm behaviour require further investigation.

In this work, we approach this challenge by modelling the population dynamics as a complex network of interactions between individuals. By applying the Interaction Networks (INs) framework [12], we can visualise and quantify the relationships between individuals as they evolve during the optimisation process. This network-based approach allows us to capture the direct influences of local search operators on individual particles and the emergent, system-level behaviours that arise from the collective interactions of the swarm. The use of network metrics, such as interaction diversity and network portrait divergence, enables a deeper analysis of how information is shared and propagated within the swarm.

Our experimental results demonstrate the potential of this network-based approach to offer a richer, more comprehensive understanding of algorithm behaviour in complex optimisation scenarios. By integrating standard performance metrics from the optimisation field with advanced network analysis techniques, we reveal new dimensions of algorithm performance that were previously underexplored. Specifically, our findings suggest that certain combinations of local search operators can significantly improve PSO's ability to balance exploration and exploitation, leading to better overall convergence speed and solution quality. Furthermore, the network-based analysis highlights how different operator interactions influence swarm dynamics, offering valuable insights for the design of more effective algorithms.

The remainder of this paper is organised as follows: Sect. 2 describes the relevant technical background. Section 3 presents the experimental setup and results, discussing the main findings. Lastly, Sect. 4 summarises the main findings and discusses the limitations and opportunities for future research.

2 Theoretical Background

2.1 Particle Swarm Optimisation

Particle Swarm Optimisation (PSO) is a population-based optimisation algorithm inspired by the social behaviour of bird flocks and fish schools [4]. In PSO, a group of particles (i.e., candidate solutions) moves through the search space, adjusting their positions based on both their own previous experiences and the experiences of neighbouring particles. Each particle has a velocity that directs its movement and is updated according to the particle's best-known position (local best) and the best-known position of its neighbours (global best or other topologies).

The definition of the set of neighbouring particles, also known as communication topology, is of elevated importance as it helps balance convergence speed (how fast the population converges to similar solutions), exploration (searching new areas of the space), and exploitation (refining known good areas). This work will explore three topologies: global best, local best, and Von Neumann [9].

The Global Best Topology (i.e., the fully connected topology) is the simplest communication structure, where every particle in the swarm can access the best position found by the entire swarm. This configuration allows for fast convergence since all particles share information about the best global solution, leading them quickly towards promising areas of the search space. However, this topology may suffer from premature convergence, as particles can become trapped in local optima if they focus excessively on the global best solution and fail to explore other regions effectively.

The Local Best Topology restricts communication to a subset of particles (often the immediate neighbours in a predefined structure, such as a ring). Each particle adjusts its velocity based on its own best position and the best position found by its local neighbourhood, rather than the entire swarm. This approach promotes greater diversity in the search process by limiting the influence of a global leader, thereby reducing the risk of premature convergence. However, the convergence rate can be slower as information diffuses more gradually across the swarm.

The Von Neumann Topology is a compromise between the local and global topologies, where particles are arranged in a grid-like structure, typically with each particle interacting with its four neighbours (north, south, east, and west). This structure balances exploration and exploitation, as information spreads more efficiently than in the local topology but with more diversity than in the global topology. This topology tends to strike a good balance between convergence speed and the ability to avoid local optima, making it a popular choice in many PSO applications.

We selected these topologies because each plays a different role in balancing the trade-off between exploration and exploitation in PSO, with global best favouring faster convergence, local best promoting diversity, and Von Neumann achieving an intermediate balance. The selected memetic operators are described next.

2.2 Memetic Operators

Memetic operators in optimisation are mechanisms used within memetic algorithms, which are hybrid optimisation techniques combining global search strategies (such as population-based algorithms like the PSO) with local search methods. These operators aim to enhance the efficiency and effectiveness of the search process by exploiting both exploration (global search) and exploitation (local search) capabilities [10].

The term memetic refers to the idea of memes, which are analogous to genes in the context of evolution but represent units of knowledge or cultural information that can evolve through imitation and adaptation. Similarly, memetic algorithms apply local refinement to solutions, akin to how individuals improve through learning and adaptation. To study the impact of different local search operators in the PSO behaviour, we selected three memetic variants: PSO-PS, PSO-HC, and PSO-SA.

PSO-PS (PSO with Pattern Search) [2]: The integration with Pattern Search, a direct search method that does not rely on gradient information, enhances the local search capability of PSO by refining particle positions after each iteration. This hybrid approach allows PSO-PS to balance exploration and exploitation, where PSO performs a global search, and the PS refines solutions locally to improve convergence speed.

PSO-HC (PSO with Hill Climbing) [7]: Hill Climbing iteratively improves a single solution by incrementally exploring its neighbourhood for better solutions. In PSO-HC, the global search power of PSO is complemented by the local refinement capabilities of Hill Climbing. After each PSO iteration, HC is applied to selected particles to explore their local surroundings for improvements. This hybridisation aims to reduce the likelihood of premature convergence in PSO by using HC to fine-tune solutions, making the algorithm more effective in avoiding local optima while accelerating convergence.

PSO-SA (PSO with Simulated Annealing) [20]: Simulated Annealing is a probabilistic technique inspired by the annealing process in metallurgy, which helps escape local optima by accepting worse solutions with a controlled probability based on a temperature parameter. In PSO-SA, SA is applied to selected particles to allow for occasional uphill moves, promoting exploration and diversity within the population. The combination of PSO's global search strategy and SA's probabilistic escape mechanism enables the hybrid algorithm to explore the solution space more effectively and maintain diversity, resulting in improved performance in complex optimisation landscapes.

Understanding interactions among individuals is essential in population-based optimisation algorithms, such as PSO and its variants. In the next section, we explain how we capture the population's interactions into a network.

2.3 Interaction Networks

Interaction Networks offer valuable insights into the collective behaviour of individuals, shedding light on the role of interactions in the search for optimal solutions. By studying these interaction networks, a deeper comprehension of PSO's functioning can be achieved, enabling further exploration of its potential in solving complex problems.

In 2014, Oliveira et al. proposed a network-based approach to represent and visualise these interactions [11]. This approach is grounded in principles that allow the representation and visualisation of interactions among swarm individuals, capturing the collaborative and competitive relationships during the optimisation process.

The creation of the networks can be summarised in three steps: first, we map all the interactions that occur at the population level. Next, for each interaction mapped, we convert the individuals involved into nodes, and the link between them has a weight corresponding to their distance in the search space. Lastly, we apply a time window concept to group the networks of consecutive iterations, capturing more information in a single network. These steps are illustrated in Fig. 1.

Previously, Interaction Networks (INs) were applied to evaluate a memetic variation of PSO, known as PSO with Pattern Search. This research employs the same network creation methodology and edge weight representation presented by Santana et al. [16]. In this work, we extend the number of memetic variations of PSO and investigate the impact of different communication topologies on the algorithm's behaviour.

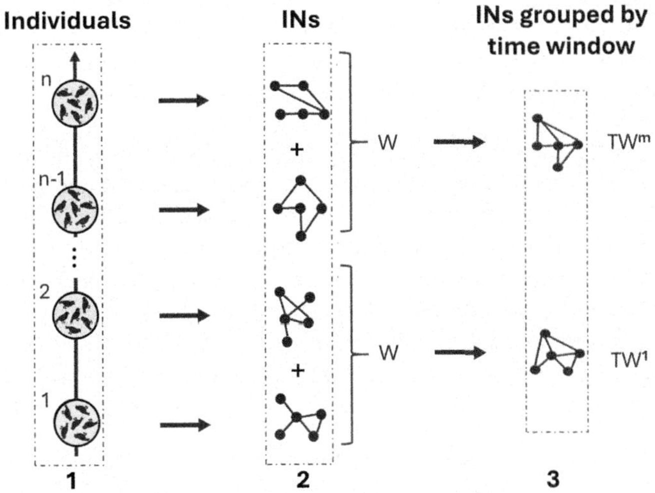

Fig. 1 Steps to create the interaction networks

2.4 Assessment Metrics

Besides the traditional analysis of the nodes' degrees and edges' weight, we employ two additional metrics to gauge the similarities between two networks (i.e., the Portrait Divergence metric) and to measure the diversity of interaction patterns within the population (i.e., interaction diversity).

The Portrait Divergence (PD) is a metric developed by Bagrow and Bollt [1] to quantify structural differences between interaction networks. This metric calculates the Jensen-Shannon divergence between the portraits, also known as B-matrices, of the networks. These portraits are derived from the information about nodes and edges in the networks, providing an encoding of their structure. The PD metric uses Jensen-Shannon divergence to measure the distance between two portraits.

PD generates a value ranging from zero to one, indicating the degree of similarity or difference between the compared networks. When applied to different swarm-based algorithms, PD can reveal whether the interaction networks exhibit similar or distinct structures, offering deeper insights into the specific characteristics of each algorithm.

The PD metric was selected due to its high flexibility, enabling the comparison of networks with different types and topologies, even when they are not defined over the same sets of nodes [5, 8, 16]. This flexibility offers a valuable advantage in analysing the interaction networks of different algorithms at various stages of execution.

Next, Interaction Diversity (ID) is a metric proposed by Oliveira et al. [12] to measure the diversity of information flow within the swarm. ID is directly related to the spatial distribution of solutions in the swarm. The more diverse the information flow, the more interconnected the nodes in the interaction graph. ID is calculated by removing a fraction of the weakest nodes (those with lower weights) from the influence graph and observing how the graph divides into isolated subgraphs [16]. The ID index varies from 0 to 1, where values close to 1 indicate high diversity in the interaction patterns.

ID plays a crucial role in efficiently searching for optimal solutions in PSO. Analysing interaction diversity makes it possible to detect stagnation patterns, assess the balance between exploration and exploitation, and compare different swarm topologies. Previous studies have used the ID metric to analyse the stagnation phenomenon in different communication topologies for PSO [12], to investigate the exploration-exploitation balance in PSO [13], and to compare exploration and exploitation in various swarm-based algorithms [14]. This work will employ the interaction diversity metric to analyse exploration and exploitation in the selected memetic algorithms.

3 Experiments and Results

In this work, we selected memetic variants of PSO as a case study to demonstrate the capabilities of interaction networks. We conducted a comparative analysis on three memetic variations of the PSO algorithm: PSO-PS [2], PSO-HC [7], and PSO-SA [20]. These algorithms were combined with three different PSO topologies: Global (PSO), Local (LPSO), and Von Neumann (VNPSO).

The topologies were implemented using the same basic structure of PSO, and the parameters were configured based on the work of Santana et al. [16]. The experiment was set up with 30 runs of each algorithm, with a population of 100 individuals and 500 iterations as the stopping criterion for PSO. A linear reduction of the inertia factor from 0.9 to 0.4 was applied, while the cognitive and social coefficients were set to 1.496. The objective function dimensions were set to 50 [19], well-known benchmarking problems used to evaluate algorithm performance, were selected. This parametrisation was applied to all algorithms and their topological variations.

For the Hill Climbing-based algorithms, we used a vector length parameter equal to the number of dimensions (50), initialised with the value 1 to represent the initial step size. The acceleration was set to 1.2 to adjust the step size in different directions. The stopping criterion was the difference between the fitness values before and after execution, being <0.0001.

In the case of the Simulated Annealing-based algorithm (PSO-SA), the maximum (initial) temperature was set to 100 and the minimum temperature to 0. The stopping criterion was 500 iterations or reaching the minimum temperature. A damping factor of 1 was applied. Lastly, for the Pattern Search-based algorithms, the "radius" and initial "delta" parameters were set to 2.0 and 1.0, respectively.

Each algorithm execution, comprising 500 iterations, was divided into 50 groups of 10 iterations. Therefore, 50 time windows (TW) were obtained for each algorithm, numbered from 1 to 50. The parameter values used align with previous similar works [6, 16, 18]. Next, the adjacency matrices of each group were summed, resulting in a single matrix that encodes the interaction patterns within a given time window.

We divided our analysis into three main sections, designed to analyse the overall characteristics of the networks generated, the similarities and differences between the networks of different memetic operators and topologies, and the behaviour of the interaction dynamics of the populations.

3.1 Network Characterisation

One of the first steps in characterising the INs is identifying strongly connected components and analysing the distribution of influence among the nodes in the interaction networks. These components represent hubs of information exchange within the network, and the relationship between the number of these components and the

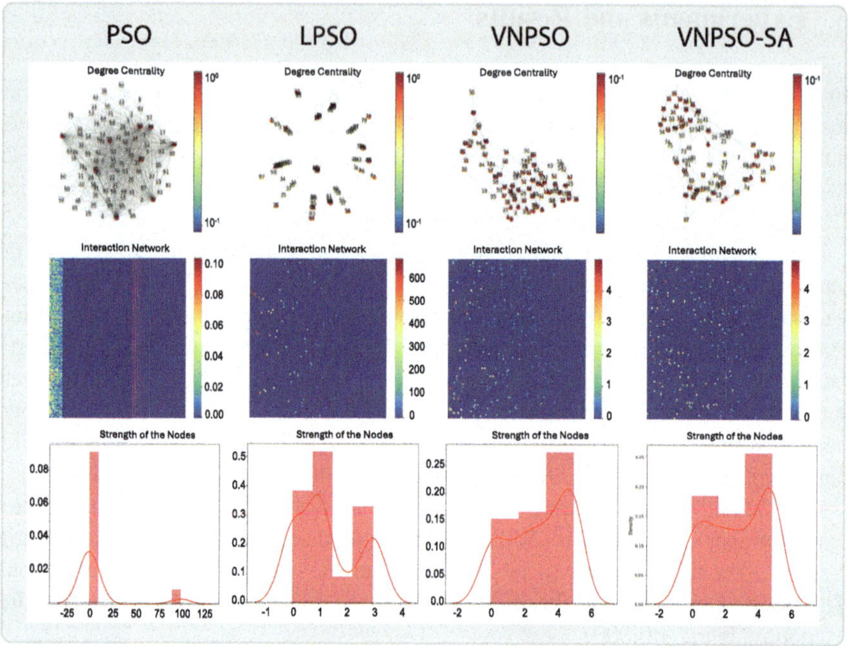

Fig. 2 Examples of networks generated in the Rosenbrock function, shown in the following order (top-down, left-right): PSO, LPSO, VNPSO, and VNPSO-SA

connections within and between them provides a strong indication of communication diversity. Furthermore, the strength of the nodes indicates the presence of influential individuals in the swarm, which is also related to the diversity of information flow [16]. The results in Fig. 2 show examples of networks for each algorithm assessed.

As observed in the first row of Fig. 2, the global best topology produces the densest networks in terms of the number of links. In this topology, the population is connected to a single influential individual at each iteration, causing the entire population to be influenced by this leader and move towards it. In the global topologies (Fig. 2, first row), we see the presence of multiple hubs, representing the leaders of the population in a given time window. Generally, a low number of hubs is associated with exploitation behaviour, while a higher number of hubs is linked to exploratory behaviour. Another finding was that the memetic phase introduced in the algorithms did not affect the interaction patterns, just the convergence speed. Comparing the VNPSO and VNPSO-SA, one can notice similar characteristics of their networks.

In contrast, the local topology (second row of Fig. 2) displays a markedly different connection pattern. Here, each individual is connected only to its two predefined neighbours, leading the networks to adopt a more linear or even cyclic structure, where hubs are not as prominent as in the global topology.

Lastly, the Von Neumann topology, as expected, exhibits behaviour that lies between the global and local topologies. In the third row of Fig. 2, we see that the Von Neumann topology forms subnetworks or clusters (representing the distinct neighbourhoods). Notably, the memetic aspect does not significantly alter the connection patterns of PSO; instead, the most pronounced differences arise from the communication topologies. However, the memetic components are expected to play a notable role in shaping the weights of the links in the INs.

3.2 Comparison of Interaction Networks

In this section, we compare the behaviour of the algorithms at different stages of the optimisation process to identify differences between the topologies and memetic operators. In Fig. 3, each pixel represents the PD score when comparing network structures across time windows for different operators. The x and y axes range from the time window 1–50, and PD values close to 0, indicating that the networks' structures are similar (i.e., interactions between the agents). The first column compares each algorithm with itself; this is the only symmetrical heatmap and provides a baseline behavior for each evaluated topology. The subsequent columns illustrate the differences between memetic operators and this baseline.

As shown in the first row of Fig. 3, both PSO-PS and PSO-HC exhibit significant similarity to the standard PSO in terms of convergence patterns and the distribution of points across time windows, indicating high similarity. However, PSO-SA demonstrates a different distribution in specific time windows, with a bottleneck around TW16 and a shift in the direction of the most similar points, suggesting a potential variation in the search for optimal solutions. This difference can be attributed to the application of Simulated Annealing, which introduces a probabilistic approach to optimisation.

In the analysis comparing LPSO (Local PSO) with its memetic variations, consistent patterns were observed across the graphs. When comparing LPSO with itself, there is a predominance of points in the central region of the graph, with subtle variations in the upper left and lower right areas. These variations indicate a similar trend of convergence across independent runs. When analysing LPSO combined with Pattern Search, the pattern suggests improved solutions over time, implying that this combination may lead to faster convergence and enhanced optimisation performance.

A consistent pattern of convergence and similarity is also observed when comparing VNPSO (last row of Fig. 3) with its memetic variations. Most graphs display a convergence region in the lower left, indicating a trend towards similar, closely related solutions across the algorithms. Additionally, a band along the diagonal suggests gradual convergence over time, with consistent progress towards optimal solutions.

However, a variation in behaviour is noted when analysing VNPSO-PS specifically (last row of Fig. 3). Here, the convergence band is more inclined and reaches the top of the graph later in the execution, a difference attributed to the introduction

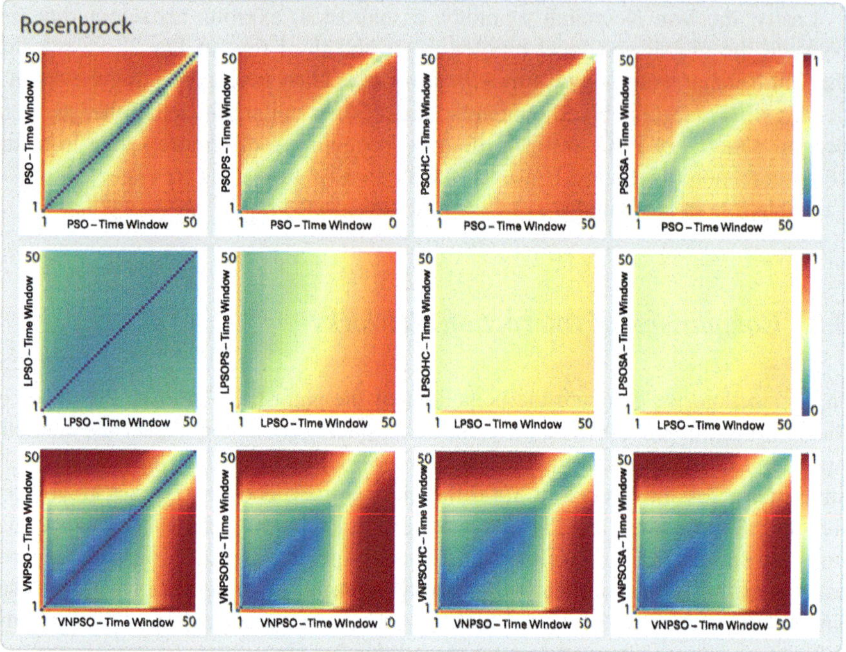

Fig. 3 Comparison of PD values between networks of PSO, PSO-PS, PSO-HC, PSO-SA, LPSO, LPSO-PS, LPSO-HC, LPSO-SA, VNPSO, VNPSO-PS, VNPSO-HC, and VNPSO-SA in the Rosenbrock function

of the Pattern Search local search, which intensifies exploration in certain regions of the search space. These insights highlight the nuances of convergence and the impact of memetic variations on VNPSO dynamics.

Since its inception, PSO has been affected by the problem of premature stagnation, resulting in convergence before reaching optimal solutions [12]. A key factor contributing to this stagnation is the lack of spatial diversity within the swarm, leading to a limited search of the solution space. This spatial diversity emerges in PSO from the social interactions among particles. In the next section, we employ the Interaction Diversity metric to analyse the diversity of the assessed PSO variations.

3.3 *Interaction Diversity*

The analysis of interactions in the PSO, PSO-PS, PSO-HC, and PSO-SA algorithms, all using the Global topology over 50 time windows, revealed distinct patterns, as shown in the first row of Fig. 4. In the case of the classic PSO, a sharp drop in Interaction Diversity was observed in the initial stages, followed by a gradual recovery in subsequent time windows. This behaviour reflects a highly exploratory search early

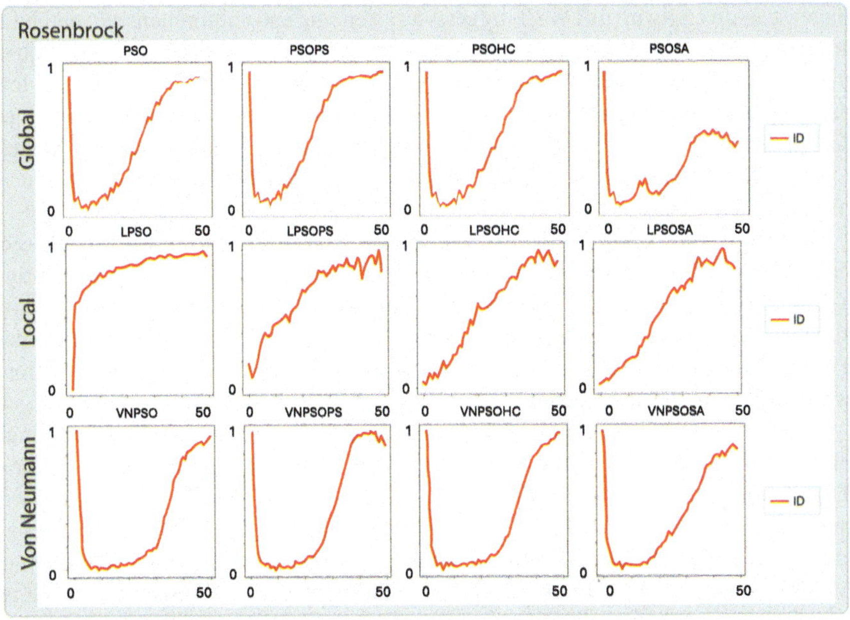

Fig. 4 Interaction diversity metric of algorithms in the Rosenbrock function

in the optimisation process, followed by a phase of low diversity (exploitation), indicating rapid convergence of the swarm to a potential local minimum. Subsequently, an increase in diversity signals renewed exploration of promising areas.

Despite the incorporation of local search in the memetic algorithms, both PSO-PS and PSO-HC exhibit similar behaviours (first row of Fig. 4, second and third plots), maintaining relatively stable Interaction Diversity (ID) throughout the execution, with only minor fluctuations. In contrast, PSO-SA (first row of Fig. 4, last plot) shows a steep decline in ID during the initial time windows, followed by a low, steady curve with subtle fluctuations across the execution period. The Simulated Annealing strategy introduces a randomised element, enabling the swarm to escape local minima. This behaviour may significantly reduce ID, suggesting that the swarm is confined to a low-diversity search region.

When analysing the local topology, distinct behaviours are observed in the LPSO, LPSO-PS, LPSO-HC, and LPSO-SA graphs (second row of Fig. 4). LPSO's rapid transition from low to high diversity at the outset suggests an intense exploration phase where the swarm extensively searches the solution space. In LPSO-PS, the initial high diversity indicates broad exploration at the beginning, followed by a drop in diversity until TW15, suggesting that the swarm has identified promising regions and is concentrating on exploiting them. The subsequent increase in diversity indicates a return to broader exploration, possibly in search of alternative solutions.

In LPSO-HC (second row of Fig. 4, third plot), a gradual progression from low to high diversity is observed, indicating a balanced exploration strategy. The steady increase in diversity up to TW30 suggests a systematic and comprehensive search for solutions. The sharp fluctuations and slight decline afterward may represent adjustments and refinements in the search for better solutions. Conversely, the initial low diversity in LPSO-SA (second row of Fig. 4, fourth plot) indicates a constrained search focused on specific regions. The oscillations along the main diagonal suggest a balance between exploration and exploitation, allowing the swarm to escape local minima.

In the VNPSO, VNPSO-PS, and VNPSO-HC algorithms (last row of Fig. 4, second and third plots), an initial phase of high diversity is followed by a drop, suggesting a shift towards exploiting promising regions. A subsequent sharp increase in diversity signals a new phase of broad exploration, indicating a balanced approach between exploration and exploitation that enables the swarm to discover promising solutions while exploring the search space.

For VNPSO-SA (last row of Fig. 4, fourth plot), the algorithm begins with high diversity but soon experiences a sharp drop, indicating a focus on specific regions. The sustained low diversity suggests intense exploitation, possibly leading to local minima. The later increase in diversity towards the end of the execution indicates an attempt to escape these local minima and explore alternative solutions in pursuit of a global optimum. This dynamic highlights the influence of Simulated Annealing, which facilitates the swarm's ability to avoid local minima.

4 Conclusion

The network-based evaluation presented in this study highlights the significance of swarm interaction dynamics in determining the performance of memetic swarm algorithms. Using Interaction Networks, we visualised and quantified how different memetic operators influence the structural behaviour of the swarm across various topologies. PSO-PS and PSO-HC demonstrated strong similarity to the standard PSO, particularly in the consistency of their convergence patterns.

In contrast, PSO-SA fostered greater network diversity, enhancing exploration but occasionally delaying optimal convergence. The application of Portrait Divergence and Interaction Diversity metrics provided a deeper understanding of how communication topologies affect swarm cohesion and search efficiency. Although our analysis focuses on memetic variants of the PSO, we emphasise that the methodology applied here can be extended to analyse other population-based algorithms. Future work could expand this network-based analysis to other techniques, further exploring the relationship between interaction patterns and optimisation performance.

References

1. Bagrow, J.P., Bollt, E.M.: An information-theoretic, all-scales approach to comparing networks. Appl. Netw. Sci. **4**(1), 45 (2019). https://doi.org/10.1007/s41109-019-0156-x
2. Bao, Y., Hu, Z., Xiong, T.: A PSO and pattern search based memetic algorithm for SVMs parameters optimization. Neurocomputing **117**, 98–106 (2013). https://www.sciencedirect.com/science/article/pii/S0925231213002038
3. Chen, J., Qin, Z., Liu, Y., Lu, J.: Particle swarm optimization with local search. In: 2005 International Conference on Neural Networks and Brain, vol. 1, pp. 481–484. IEEE (2005)
4. Eberhart, R., Kennedy, J.: Particle swarm optimization. In: Proceedings of the IEEE International Conference on Neural Networks, vol. 4, pp. 1942–1948 (1995)
5. Faskowitz, J., Puxeddu, M.G., van den Heuvel, M.P., Mišić, B., Yovel, Y., Assaf, Y., Betzel, R.F., Sporns, O.: Connectome topology of mammalian brains and its relationship to taxonomy and phylogeny. Front. Neurosci. **16** (2023). https://doi.org/10.3389/fnins.2022.1044372. https://www.frontiersin.org/articles/10.3389/fnins.2022.1044372
6. Guimaraes, F.V.C.: Anàlise do impacto de operadores locais no algoritmo pso memético. XI, 41 f.: il. ; 29 cm (2022)
7. Lim, A., Lin, J., Xiao, F.: Particle swarm optimization and hill climbing for the bandwidth minimization problem. Appl. Intell. **26**(3), 175–182 (2007). https://doi.org/10.1007/s10489-006-0019-x.
8. Liu, W., Xie, J., Zhang, C., Yamada, M., Zheng, N., Qian, H.: Robust graph dictionary learning. In: The Eleventh International Conference on Learning Representations (2023). https://openreview.net/forum?id=qxRscesArBZ
9. Mendes, R.: Neighborhood topologies in fully informed and best-of-neighborhood particle swarms. IEEE Trans. Syst. Man Cybern. Part C: Appl. Rev. **36**(4), 515–519 (2006). https://doi.org/10.1109/TSMCC.2006.875410
10. Moscato, P., Norman, M.G.: A memetic approach for the traveling salesman problem implementation of a computational ecology for combinatorial optimization on message-passing systems. Parallel Comput. Transputer Appl. **28**(1), 177–186 (1992). http://citeseerx.ist.psu.edu/viewdoc/summary?doi=10.1.1.50.1940
11. Oliveira, M., Bastos-Filho, C.J., Menezes, R.: Towards a network-based approach to analyze particle swarm optimizers. In: IEEE SSCI 2014—2014 IEEE Symposium Series on Computational Intelligence—SIS 2014: 2014 IEEE Symposium on Swarm Intelligence, Proceedings, pp. 166–173 (2015). https://doi.org/10.1109/SIS.2014.7011791
12. Oliveira, M., Pinheiro, D., Andrade, B., Bastos-Filho, C., Menezes, R.: Communication diversity in particle swarm optimizers. In: Dorigo, M., Birattari, M., Li, X., López-Ibáñez, M., Ohkura, K., Pinciroli, C., Stützle, T. (eds.) Lecture Notes in Computer Science, vol. 9882. LNCS, pp. 77–88. Springer International Publishing, Cham (2016). https://doi.org/10.1007/978-3-319-44427-7_7
13. Oliveira, M., Pinheiro, D., MacEdo, M., Bastos-Filho, C., Menezes, R.: Better exploration-exploitation pace, better swarm: examining the social interactions. In: 2017 IEEE Latin American Conference on Computational Intelligence, LA-CCI 2017—Proceedings, vol. 2017, November, pp. 1–6 (2018). https://doi.org/10.1109/LA-CCI.2017.8285712
14. Oliveira, M., Pinheiro, D., Macedo, M., Bastos-Filho, C., Menezes, R.: Uncovering the Social Interaction in Swarm Intelligence with Network Science, pp. 1–11 (2018). http://arxiv.org/abs/1811.03539
15. Peer, E.S., van den Bergh, F., Engelbrecht, A.P.: Using neighbourhoods with the guaranteed convergence PSO. In: Proceedings of the 2003 IEEE Swarm Intelligence Symposium. SIS'03 (Cat. No. 03EX706), pp. 235–242. IEEE (2003)
16. Santana, C., Oliveira, M., Bastos-Filho, C., Menezes, R.: Beyond exploitation: measuring the impact of local search in swarm-based memetic algorithms through the interactions of individuals in the population. Swarm Evol. Comput. **70**, 101,040 (2022). https://doi.org/10.1016/j.swevo.2022.101040

17. Shi, Y., Eberhart, R.: A modified particle swarm optimizer. In: 1998 IEEE International Conference on Evolutionary Computation Proceedings. IEEE World Congress on Computational Intelligence (Cat. No.98TH8360), pp. 69–73 (1998). https://doi.org/10.1109/ICEC.1998.699146
18. Silva, C., Santana, C., Bastos-Filho, C.: Impact assessment of memetic operators on communication topologies for the PSO. Technical report
19. Wu, G., Mallipeddi, R., Suganthan, P.: Problem definitions and evaluation criteria for the CEC 2017 competition and special session on constrained single objective real-parameter optimization. Nanyang Technological University, Singapore, Technical Report, pp. 1–18 (2016)
20. Yang, H., Yang, Y., Yang, Z., Zhang, L.: An improved particle swarm optimization algorithm based on simulated annealing. In: 2014 10th International Conference on Natural Computation (ICNC), pp. 529–533 (2014). https://doi.org/10.1109/ICNC.2014.6975891
21. Yang, X.S.: Nature-Inspired Metaheuristic Algorithms. Luniver Press (2010)

A Hybrid Framework for Quantifying and Analyzing the Structural Properties of Human Retinal Vessel Networks

Hitalo Silva, Diego Silva, Carmelo Bastos-Filho, Alexandre Rosa, Rafael Albuquerque, Arlington Rodrigues, Luigi Tahara, Luiz Roisman, and Samuel Moscavitch

Abstract This paper presents an automated approach for analyzing retinal vascular networks using advanced graph-based techniques. The proposed method integrates deep learning models for the segmentation of retinal vessels, generation of surface meshes, and network construction to quantify and cluster retinal features. A key component of the framework is the transformation of retinal vessels into a biomedical network, where topological and semantic graphs are constructed to capture detailed vascular structures. Network metrics, including clustering coefficients, centrality measures, and routing efficiency, are extracted to characterize vascular networks. Then, Graph Neural Networks (GNN) models are applied to encode each retinal vascular network as a vector of numbers; a similar vector means a similar structure. Finally, clustering algorithms group analogous patterns and identify potential anomalies. Preliminary results, based on a dataset of **1221** retinal augmented surface meshes, demonstrate the effectiveness of the approach in distinguishing groups with similar characteristics, highlighting the potential for the early detection of diseases such as diabetes, hypertension, and cardiovascular conditions.

Keywords Biological networks · Network science · Computational intelligence · Computer vision

H. Silva (✉)
CCES, IFPE, Recife, PE, Brazil
e-mail: hos@ecomp.poli.br

H. Silva · C. Bastos-Filho · R. Albuquerque · A. Rodrigues
ECOMP, Universidade de Pernambuco, Recife, PE, Brazil

D. Silva
Universidade Católica de Pernambuco, Recife, PE, Brazil

A. Rosa · L. Tahara · L. Roisman
ICM, Universidade Federal do Pará, Belém, PA, Brazil

S. Moscavitch
Universidade Federal Fluminense, Niterói, RJ, Brazil

M. Macedo et al. (eds.), *Complex Networks XVI*, Springer Proceedings in Complexity,
https://doi.org/10.1007/978-3-031-93619-7_4

1 Introduction

The human body contains a **complex network** of arteries and veins spanning kilometers, enabling the seamless exchange of substances among cells, tissues, and organs. The proper functioning of the vascular system is vital for a healthy and enduring lifespan. Many diseases can emerge as a result of problems with blood flow [1].

Vision is the most dominant sense in the human body and is essential for daily activities. Difficulty in seeing might transform simpler tasks, such as walking and writing, into challenging ones. Among the structures of the eye, the retina is one of the most important. Retinal diseases are the main reasons for visual impairment and blindness and are usually related to systemic diseases, such as diabetes, hypertension, cardiovascular illness, autoimmune disorders, and infectious processes [2].

Overall, the retina offers a unique window to the microcirculation system via a non-invasive method, such as fundus retinography. In addition to previously mentioned alterations, arteriolar narrowing can predict the progression of chronic kidney disease (CKD) and is associated with an increased risk of coronary artery disease (CAD) and stroke. In that way, it is essential to evaluate the fundus exam to detect and assess several systemic diseases [2, 3].

Artificial intelligence (AI)-based solutions are widely used in the screening and assessment of retinal diseases and structure, with studies showing reliability comparable to that of experienced examiners. The application of deep learning (DL) in screening increases efficiency and may enhance the accuracy of the diagnosis, reduced analysis time, less variability intra or interobserver, and scalability. Additionally, the capacity to detect small nuanced changes and patterns, not easily noticeable per human, can enable early detection of systemic comorbidities and stratify their stages [3].

It is necessary to create objective metrics that represent the structure and characteristics of the retina, which can be efficiently stored and compared over time. Approaches to quantifying such metrics are being developed to minimize observer errors. Furthermore, these methods are expected to enable predictions and inferences based on retinography analyses. Despite advances, current approaches have limitations in capturing physiological characteristics, such as microvessels, side branches, and bifurcations; neovascularization; distinguishing veins from arteries; and assessing hemodynamics and vasomotricity, all of which are technically challenging to measure [4]. Another relevant aspect that might help reduce complications, including vision impairment or blindness caused by retinal diseases, is monitoring the patient's clinical condition over time. Some hospitals already store clinical data and produce comparative and historical reports, however, there are no standardized methods for exam interpretation and formatting [2].

Segmentation of blood vessels in retinal images is essential for the prevention, diagnosis, and evaluation of ocular diseases, as these conditions often alter the vascular morphology of the retina. Manual segmentation is challenging due to low contrast, curvilinear structures, and variable illumination, leading to inconsistent results among clinicians. To address this, deep learning frameworks have been developed to automatically capture microvessels and extract vascular features, significantly aiding diagnosis and treatment. However, currently there is no universally accepted model for segmentation of the retinal vessels. Recent trends focus on incorporating self-attention mechanisms to capture global information and reduce image information loss. This approach, combined with convolutional networks, improves segmentation efficiency and accuracy. In addition, ensemble learning techniques integrate multiple models to enhance performance [5–7].

Ravandi and Ravandi [4] proposed analyzing networks derived from coronary arteries and veins from four main perspectives: structural characteristics, distribution of connectivity levels, network integration, and controllability. The structural evaluation focuses on examining the arrangement of nodes and edges within the network. Key metrics are computed to provide insights into the networkâŁ™s structure, including the average degree of connectivity, the clustering coefficient, the diameter of the network and the number of *λ-branches*. Together, these metrics help to understand the overall topology and organizational properties of vascular networks. The *λ-branches* are characteristic structures in graphs, consisting of a parent node connected to exactly two child nodes, which in turn are not connected to any other nodes. The authors hypothesize that the graphs derived from the cardiovascular system, when associated with pathological conditions, exhibit a higher number of *λ-branches* compared to the graphs originating from healthy cases. This behavior could indicate that blood is not being adequately supplied to diseased areas. An increased presence of *λ-branches* can be interpreted as an indicator of neovascularization, a phenomenon described in the medical literature as the formation of new blood vessels in response to inadequate blood supply, as highlighted in [8]. This approach allowed, in a preliminary and **non-automated way**, to analyze the structure of the coronary arteries using graphs. It was possible to evaluate the shape, degree distribution and controllability of the complex network in a healthy and diseased coronary angiogram.

The integration of DL and Network Science offers a robust approach to analyzing complex systems, such as biomedical networks. While DL extracts intricate patterns, Network Science provides a framework to analyze relationships and structures, thereby improving interpretability. In biomedical applications, this interpretability is critical for ensuring clinical relevance. For instance, metrics like *λ-branches* can correlate with physiological phenomena such as neovascularization. Ensure results align with biological knowledge can enhance trust and usability [9].

Inspired by Ravandi and Ravandi's research, we developed and described in this paper a novel process to quantify the structure of human retina vessels automatically from retinographies by applying a hybrid solution that joins DL and Network science algorithms/techniques. We also created a new retinography database, composed of images, phenotypes, and clinical information.

2 Proposal

In order to provide more interpretability in the human retinal study and its association with systemic diseases, we developed a novel process to automatically stratify human retinas. It is employed a hybrid solution that joins DL and Network Science algorithms/techniques. Retinal vessels are extracted from retinographies and then converted to biomedical graphs (also called semantic graphs). After that, an "open box" analysis is conducted via semantic enrichment (e.g., network metrics generation) and qualitative investigation (e.g., pattern recognition and clustering similar networks). The process consists of five independent and loosely coupled steps/modules: 1—Segmentation; 2—Rendering; 3—Graph Generation; 4—Network Analysis; and 5—Qualitative Inference. The development of loosely coupled modules facilitates the maintenance and evolution of each separately. In addition, the user can execute specific modules for isolated needs.

The **segmentation module** receives a biomedical image, in this case, a retinography, and pre-processes it. After that, it extracts the desired structures and generates another image in the same format containing them. The **rendering module** receives the image containing the extracted parts and converts them to unstructured surface meshes. A surface mesh is a mathematical and computational representation of the surface of a three-dimensional (3D) object, composed of interconnected geometric elements such as vertices, edges, and faces.

The **graph generation module** transforms unstructured surface meshes into biomedical networks. In this work, the first step to transform a surface mesh into a graph involves determining the centerline of the entire structure [10]. A centerline is a geometric representation that captures the central trajectory or axis of an elongated structure or object, often serving as a simplified abstraction of its shape. Usually, an object is composed of one or more centerlines (segments). At each fork in the path, a new one is generated. Based on the centerlines, it is possible to define the nodes and edges that make up the network/graph. The number of nodes is determined by the length of each segment and the original shape of the superficial mesh in that place. Furthermore, if two consecutive centerline points have a very different radius or if there is a significant change in direction or angle between them, another node is created. The network generated—called **Topological Graph**—mimics the original structure of the segments. Each topological node stores information regarding position, diameter, length, bifurcations, and direction in its attributes. Topological graphs are relevant for the geometric visualization of networks. However, they are not suitable for analysis and metrics calculation. Thus, it is essential to develop a process for transforming a topological graph into another type capable of being analyzed, which we call **semantic graphs**. The semantic graphs defined in this project represent each topological segment with a maximum of three vertices: one or two vertices for intersegment connection and one (intrasegment) to represent the segment itself. The intrasegment node stores as attributes the average and standard deviation of the respective segment radius, geometric barycenter, tortuosity, and length. The semantic edges are weighted and hold the distance between the nodes as weight.

The **network analysis module** receives a semantic graph and processes it in two ways: network metric extraction and global and local graph analysis. The objective metrics produced in the first step are the number of nodes, edges, and λ-branch; average degree, in-degree, and out-degree; clustering coefficient; diameter and radius; neighborhood connectivity; centralities of betweenness, proximity, and eccentricity; average of the shortest paths between any two nodes; and routing efficiency of the network. In the second step, Graph Neural Networks (GNNs) are applied to the graphs to capture relevant information at both local and global levels. GNNs are specialized neural network architectures designed to operate on graph-structured data, making them particularly useful for tasks that require explicit modeling of relationships, such as node classification, link prediction, whole-graph classification, subgraph recognition, and graph clustering [11]. When labels are available, GNNs can be trained in a supervised manner, using them to guide learning and enable class prediction during inference. However, in many cases, labels are unavailable or incomplete, making supervised learning infeasible. To address this, Self-Supervised Learning (SSL) emerges as a potential solution [11].

SSL is a machine learning technique that creates data representations without needing labeled data. Authors organize SSL into two main categories: contrastive and predictive. In the contrastive models used in this work, an encoder is implemented in the learning process to compare different views of the data. This encoder employs an objective function to maximize the similarity between views generated from the same original data (positive pairs) and minimize the similarity, increasing the dissimilarity, between views of different data (negative pairs). The contrastive model also utilizes auxiliary tasks, known as pretext tasks, which are formulated to encourage the model to learn representations that can distinguish well between different inputs. After training, generalized models are generated and can be applied to subsequent tasks, known as downstream tasks [11].

The **qualitative inference module** is designed to analyze and interpret complex datasets to identify meaningful patterns that could indicate a range of conditions or phenomena. These include the presence of diseases, the detection of anomalies, assessment of healthy states, signs of aging, clustering trends, or regions of neovascularization. The data analyzed by this module is often diverse, comprising information collected from one or more patients over a period of time. Typical inputs include biomedical graphs, which map biological interactions; network metrics, representing quantitative characteristics of networks; and phenotypic data, detailing observable patient traits and symptoms. By integrating and analyzing these heterogeneous data sources, the module aims to provide comprehensive insights into patient health, assist in early diagnosis, and support monitoring disease progression or recovery.

3 Methodology

The development of the proposed process followed a hybrid approach that incorporated solutions tailored to each module's requirements. This multidisciplinary approach ensures that each module operated efficiently and contributed to the overall effectiveness of the process.

One of the most critical factors in the success of deep learning (DL) models is the quality and diversity of the data. To ensure more reliable and robust results, we used retinal images from three different datasets. The first dataset, Digital Retinal Images for Vessel Extraction (DRIVE), consists of 40 retinal images and the corresponding masks (black-and-white images representing vessels and the fundus). We used 20 images to train and 20 images to validate the segmentation model. The entire DRIVE dataset was also applied as input aiming to validate the proposal. The second dataset, STructured Analysis of the Retina (STARE), contains 400 retinal images, of which only 40 are labeled with blood vessel segmentation masks, 20 were used for training and 20 were reserved to validate the proposal. Finally, we created a custom dataset comprising 888 retinal images of healthy retinas, along with clinical information such as gender, age, date of birth, eye diseases, systemic diseases, date of exam, and laterality. Initially, only 35 retinas were used to validate the proposed method. In total, 60 images were used for training the segmentation model and 95 images were used during validation. This diverse set of retinal images, combined with clinical data, ensures a more comprehensive evaluation of the segmentation model and enhances its ability to generalize across different scenarios. In addition, it enables us to search for associations between retinal networks structures/patterns and eye and systemic diseases.

To segment retinal vessels from fundus images, we developed a novel model and a custom loss function. The model integrates elements from the architectures proposed in Ronneberger et al. [5], Peng et al. [7], and Cui et al. [6], alongside the adapted Skeleton Recall Loss approach introduced by Kirchhoff et al. [12]. The process begins with dividing the dataset into training and testing sets. Subsequently, the images undergo pre-processing to enhance their quality and ensure consistency, making them suitable for analysis. Key preprocessing steps include dataset normalization, CLAHE (Contrast Limited Adaptive Histogram Equalization), and Gamma Correction. These techniques work together to improve image quality and standardize the dataset, ultimately enabling more easily distinct retinal fundus of vessels. After that, the images are sampled into 200,000 training patches of size 48x48 pixels, which are further split into training and validation sets. The following parameters were used during the training of the segmentation model: Epochs: 500, Batch Size: 384, Learning Rate: Ranges from 0.001 to 0.00001, Optimizer: Adam, Gradient Clipping: 1.0. In this work a novel loss function, TverskyLossBinSigmoidSkel, was introduced. This loss function incorporates a new weight parameter (θ) and divides the mask between thicker vessels and thinner structures, referred to as "skeletons". The parameter θ adjusts the loss contribution from the skeleton vessels, helping to mitigate the errors caused by their thin structure and improving the overall segmentation performance.

The surface mesh module converts an image into a 3D surface mesh. The first step involves transforming the pixels that represent the retinal vessels into 2D vectors, which are composed of nodes and links. This process was accomplished using the Potrace Python library. Once the 2D vectors were generated, they were then converted into 3D vectors using the Trimesh library. The resulting 3D surface meshes were subsequently stored as STL files, allowing further analysis and visualization. Each surface mesh was augmented by rotating it 15 °C at a time, from 0 to 180 °C. This approach aims to assess whether the process can generate similar encoding for the same retina, even when subjected to slight modifications due to rotation. Thus, based on 95 retinas, 1221 surface meshes were created. Some rotations led to errors during the graph generation process, probably due to distortions in the mesh structure that interfered with the accurate creation of the graph.

The process of converting a surface mesh into a biomedical graph begins with the extraction of centerlines. This step was carried out using an adapted version of the VMTK library, with the parameter "Decimation Aggressiveness" set to 4.0 (default is 3.0, range of values 0.0–15.0). This parameter controls how closely the algorithm follows the original structure of the surface mesh during centerline extraction, balancing fidelity and computational efficiency. The more complex your surface mesh (number of nodes, links, bifurcations, and vessel endings; level of tortuosity; the presence of degenerated faces; etc.), the more complex the centerline extraction process will be. Usually, during the extraction, the algorithm establishes one source and many targets (in this case, vessels endings). The source point is located at the northwestern edge of the mesh, which is the default initialization method of the library. Using the extracted centerlines, topological graphs are constructed by mapping each centerline segment into nodes and edges. Although these graphs effectively capture the topological structure of the network, they are not indicated for depth analysis of network characteristics. To address this limitation, the topological graphs are further converted into semantic graphs, which include enriched information about the network's features and properties. Both, the topological and semantic graph conversions, were implemented using custom algorithms developed by the authors. This module enables the transformation of biomedical surface meshes into comprehensive graph representations suitable for advanced analysis.

The network analysis module consists of two submodules, each with distinct roles in processing and analyzing semantic graphs derived from retinal vessel networks. The first submodule is for the extraction of network metrics. This submodule focuses on computing various network metrics from graph structures, primarily using directed links. The main aim of making the graphs with direct links was to try to mimic the blood flow inside the vessels. However, it is not possible to determine the blood flow because we are still unable to differentiate veins from arteries based on the datasets accessed, and the source point, defined by the previous module, is not always located in the largest cavity of the surface network, indicating where blood can come in the retina. In addition, specific metrics, such as Routing Efficiency, require an undirected version of the graph. To address this, all links were created as undirected. Additionally, the authors introduced a specialized function to identify and mark λ-branch formations, enhancing the analysis of vascular structures. The second submodule performs pattern extraction using GNN models, uncovering local and global patterns, characteristics, and information from graphs through the application of InfoGraph and MVGRL. Both InfoGraph and MVGRL were trained using contrastive SSL techniques, with Personalized PageRank as diffusion information mode, to generate embeddings for retinal vessel networks. In addition, an investigation was conducted to verify if and which network metrics could be used as attributes of the graph nodes, aiming to help the effectiveness of the GNN models.

The **qualitative inference module**, constrained by the lack of clinical condition labels for all retinographies, presented a challenge in applying traditional supervised learning methods. To overcome this challgence, we adopted an unsupervised approach, utilizing three clustering algorithms: K-means, density-based spatial clustering of applications with noise (DBSCAN), and hierarchy DBSCAN (HDBSCAN). These algorithms were applied to cluster the encoded data generated by the GNN models, with the aim of uncovering patterns that may represent distinct conditions or underlying phenomena. To evaluate the performance and quality of the identified clusters, we employed three widely used clustering evaluation metrics: Silhouette score, Davies-Bouldin (DB) score, and Calinski-Harabasz (CH) score. The clustering results were analyzed and the configuration that yielded the best performance across these metrics was selected. This approach allowed us to derive meaningful data-driven groups that offer insight into retinal conditions and associated phenomena, even without labeled clinical data. The encoded data can either be fed directly into the clustering algorithms or undergo dimensionality reduction—via uniform-manifold approximation and projection (UMAP), Principal Component Analysis (PCA), and t-distributed stochastic neighbor embedding (t-SNE)—to enhance clustering performance by simplifying the structure of the data while retaining its essential features. After the clustering step, a normalization process is performed to ensure consistency. Specifically, the median label of the clustered retinas is applied to all augmented retinas from the same source. This ensures that all variations of a single retina, resulting from augmentations, are treated as part of the same group, preserving the integrity of the dataset for further analysis.

The authors utilized the Plotly library to visualize the topological graphs and employed Cytoscape software to render the semantic graphs.

4 Results and Discussion

This section presents and thoroughly discusses the results of extracting quantitative characteristics from biomedical networks that were generated based on retinography images. These characteristics offer valuable insights into the structural and functional properties of vascular networks in the retina, contributing to a deeper understanding of some ocular and systemic conditions.

Figure 1 visually summarizes the process of transforming a retinal image into topological and semantic graphs. Seventeen λ-branches are visible in the topological graph, highlighted with orange rectangles. Notably, two of these λ-branches

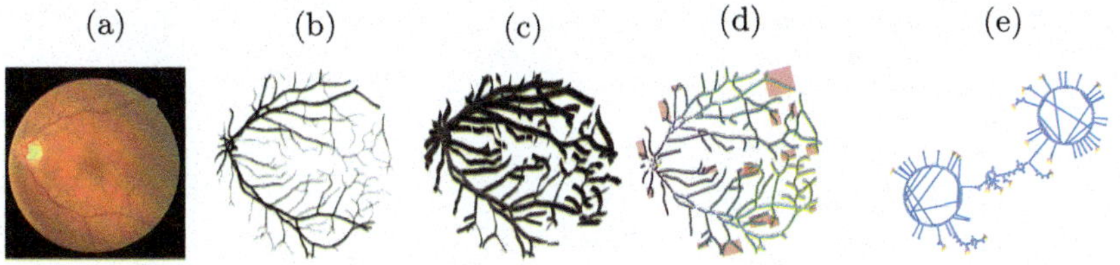

Fig. 1 Step-by-step for transforming the retinography 21 into a semantic graph. **a** Retinography. **b** Vectors representing veins and arteries. **c** Surface mesh. **d** Topological graph. **e** Semantic graph

are located very close to each other. The corresponding semantic graph, derived from the retinal image, is represented using a circular layout. In this layout, circles denote nodes, while lines represent edges. The orange circles highlight the nodes that are part of the λ-branches. The retinal vessels are generally grouped into three main regions: upper, central, and lower. The circular layout of the semantic graph reflects this structure, with two prominent circles representing the upper and lower regions, and the central area corresponding to the middle section of the retina. This arrangement mirrors the original retinal structure, making it easier to identify the λ-branches. Furthermore, this layout suggests that other segments might also belong to the same neovascularization region, offering a more intuitive way to explore potential connections within the vascular network.

An example of quantitative data extraction is described in the following lines in this paragraph. Retina number: 21, Number of Nodes: 355, Qty. Links: 380, Diameter 80, Radius 40, λ-branches: 17, Clustering Coefficient: 0.0, Closeness Centrality: 0.033, Network Density: 0.006, Betweenness Centrality 0.085, Neighborhood Connectivity: 2.679, Eccentricity Centrality: 62.656, Avg. Degree of Network Connectivity: 2.679, Avg. Shortest path length: 30.835, Routing Efficiency: 0.00933.

Each retinal network/graph is represented by a vector of numbers. This operation is called graph encoding and is executed in this work by the GNN models. If two retinal vessels are similar, the model will codify them in a similar way. Among the two GNN models tested, MVGRL demonstrated superior performance, achieving better graph differentiation, lower loss values, and higher validation scores. This highlights its effectiveness in capturing the complexity of vascular structures while improving accuracy in downstream tasks. In addition, we realize that the use of network metrics as node or link attributes reduced the effectiveness of the models. So, we set the attributes of the nodes with the values of the average and STD segment radius, curvature, and torsion; and a Boolean True if the segment is a λ-branch, otherwise False.

After conducting a comprehensive grid search that explored various configurations—including clustering algorithms, different distance metrics, and the minimum number of clusters (ranging from 2 to 15)—alongside dimensionality reduction techniques, the optimal setup was determined based on clustering metric

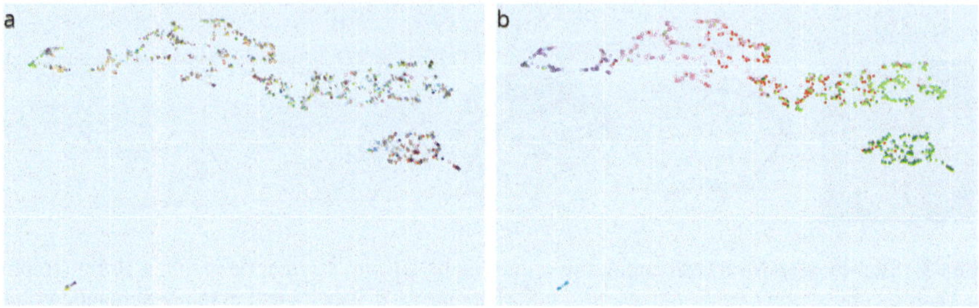

Fig. 2 Displays a total of 1221 encoded augmented retinas in a multidimensional vector. **a** before clustering, each original retina is represented by a unique symbol and color; and **b** after clustering, the retinas were grouped in six groups, each represented by a unique symbol and color

scores. The configuration combining K-means clustering, UMAP for dimensionality reduction, and six clusters emerged as the best solution, delivering the highest quality results.

Figure 2 displays a total of 1221 encoded augmented retinas in a multidimensional vector. In Subfigure (a), each original retina is represented by a unique symbol and color. The encoded data are spatially organized, with representations of the same retina placed closely, indicating that the encoding process consistently captures the essential features of each retina. This proximity suggests that the model retains key characteristics across transformations. Furthermore, Subfigure (b) reveals six distinct clusters formed within the dataset, each represented by a unique symbol and color. These clusters indicate that the model has identified underlying patterns or similarities among certain retinas, which could be associated with specific groups of features, such as the vascular structure, disease markers, or other distinguishing characteristics. In addition, some encoded data belonging to the same cluster are located at distant points, far from the main grouping. This phenomenon typically occurs when an augmented retinal mesh undergoes transformations that significantly alter its semantic graph, causing it to deviate from the characteristics of other retinas. Such variations result in slight discrepancies in the way retinal features are represented within the graph structure. Then, there are two retinas that are noticeably distanced from the main groups. These outliers may represent unusual or exceptional cases, potentially linked to unique or rare retinal conditions, or they might reflect encoding errors or anomalies in the data. More detailed analysis would be required to investigate the reasons for their separation from the larger dataset and to determine if they signify any clinical relevance.

Table 1 presents the median values of the metrics extracted from eight retinal networks, grouped by clusters. These metrics provide a summary of the key characteristics of the retinal networks within each cluster, allowing comparisons between different cluster. By analyzing the table, one can observe notable similarities and dissociation between the clusters. Some clusters exhibit comparable values for certain metrics, suggesting that they share structural or functional properties, while others show significant differences, potentially indicating distinct patterns or variations

Table 1 Displays the median value of network metrics extracted from the retinal network grouped by clusters

Cluster Id	1	2	3	4	5	6	7	8
1	22.0	54.0	8611.275	14.556	70.0	0.005	26.390	0.0639
2	16.0	48.0	8108.446	13.437	70.0	0.006	25.551	0.0689
3	24.0	52.0	9578.209	14.914	72.0	0.004	25.967	0.061
4	12.0	42.0	6270.217	8.775	60.0	0.009	20.694	0.088
5	20.0	54.0	9184.110	15.694	72.0	0.004	26.975	0.062
6	20.0	54.0	8119.117	14.223	72.0	0.006	27.195	0.066

1: Qtt of λ-branches, 2: Eccentricity, 3: Total length, 4: Total tortuosity, 5: Diameter, 6: Densite, 7: Shortest path avg, 8: Routing efficiency

in the retinal networks. A preliminary statistical analysis was performed employing the Kruskal-Wallis (KW) test to compare each distribution of network metrics across the identified clusters pair-to-pair. The study revealed that all network metric distributions in the clusters are statistically significantly different.

A preliminary result showed that the retinas contained in Cluster 2 are from patients older than 70 years, and all individuals in this cluster have been diagnosed with Hypertension (HAS). This suggests that the clustering model has successfully identified age and health-related patterns in the vascular networks, potentially linking these factors to the specific characteristics of retinal vessels. This could serve as an important marker for early identification of hypertension-related changes in retinal vasculature.

5 Conclusion and Future Works

This paper outlines the initial development of an automated system designed to objectively stratify the human retina microvascular system, employing a hybrid solution that joins DL and Network Science algorithms and techniques. The proposed process is structured into five distinct interconnected modules. The first module focuses on automatic segmentation of retina vessels from biomedical image using advanced image processing techniques. The second module generates a three-dimensional superficial mesh of the vessel network, previously represented as pixels. In the third module, the superficial mesh is converted into a biomedical graph. This graph-based representation is essential to capture the topological and structural properties of the vascular system. The fourth module leverages network science principles to extract a range of objective metrics from biomedical networks. This step enables a deeper understanding of the role of vessel structure in overall retinal health. The final module involves qualitative inference that seeks to identify meaningful patterns or anomalies in the vascular network that can indicate various health conditions.

By applying Graph Neural Networks (GNNs) and clustering algorithms, the system is able to detect potential markers of diseases such as hypertension, diabetes, and cardiovascular disorders, based on observed vascular characteristics.

Together, these modules form an integrated framework that automates the extraction, modeling, and analysis of vascular characteristics. This approach has the potential to significantly improve diagnostic accuracy, offering a more efficient and objective method to assess the health of the retina and the body. By combining advances in image processing, 3D modeling, network analysis, and machine learning, this automated system can enhance the precision of clinical decision-making and contribute to personalized healthcare strategies.

The preliminary results are promising, demonstrating the potential of our approach for analyzing retinal networks. However, there are several avenues for future work to improve and refine the methodology. First, expand the dataset with more images and clinical information. This could lead to more personalized and accurate predictions. Second, improve the algorithm that extracts the centerlines. One possible enhancement is to set the source point in the middle of the largest cavity of the surface mesh, with the aim of reducing extraction problems, as it will always be on the same place even when the mesh suffers spacial transformations. Third, a more in-depth study of how the use of network metrics in the composition of semantic graphs (present in the attributes of nodes and links) can lead to better graph-level encodings. Improve qualitative analysis maturing the GNN models and training; and clustering process. Then, conduct an extensive validation using both internal and external datasets. Increasing the interpretability of network analysis will help clinicians better understand the patterns and anomalies detected, providing more actionable insights.

References

1. WHO: Cardiovascular diseases (CVDs) (2021). https://www.who.int/news-room/fact-sheets/detail/cardiovascular-diseases-(cvds)
2. Hanssen, H., Streese, L., Vilser, W.: Retinal vessel diameters and function in cardiovascular risk and disease. Progr. Retinal Eye Res. **91**, 101095 (2022)
3. Amir Hamzah, N.A., Wan Zaki, W.M.D., Wan Abdul Halim, W.H., Mustafar, R., Saad, A.H.: Evaluating the potential of retinal photography in chronic kidney disease detection: a review. PeerJ **12**, e17786 (2024)
4. Ravandi, A., Ravandi, B.: Network-Based Approach for Modeling and Analyzing Coronary Angiography (Barbosa, H., et al., eds.). Complex Networks. Series Title: Springer Proceedings in Complexity, vol. XI, pp. 170–181. Springer International Publishing, Cham (2020)
5. Ronneberger, O., Fischer, P., Brox, T.: U-Net: Convolutional Networks for Biomedical Image Segmentation. Publication Title, Medical Image Computing and Computer-Assisted Intervention—MICCAI. Springer International Publishing, Cham (2015)
6. Cui, Z., Song, S., Qi, J.: MF2ResU-Net: a multi-feature fusion deep learning architecture for retinal blood vessel segmentation. Digit. Chin. Med. **5**, 406–418 (2022)
7. Peng, Y., Tang, Y., Luan, P., Zhang, Z., Tu, H.: MAFE-Net: retinal vessel segmentation based on a multiple attention-guided fusion mechanism and ensemble learning network. Biomed. Opt. Express **15**, 843–862. Optica Publishing Group (2024)

8. Moreno, P.R., Purushothaman, K.R., Zias, E., Sanz, J., Fuster, V.: Neovascularization in human atherosclerosis. Curr. Mol. Med. **6**, 457–477 (2006)
9. Doshi-Velez, F., Kim, B.: Towards a rigorous science of interpretable machine learning (2017). ArXiv:1702.08608
10. Wolterink, J.M., van Hamersvelt, R.W., Viergever, M.A., Leiner, T., Išgum, I.: Coronary artery centerline extraction in cardiac CT angiography using a CNN-based orientation classifier. Med. Image Anal. **51**, 46–60 (2019)
11. Xie, Y., Xu, Z., Zhang, J., Wang, Z., Ji, S.: Self-supervised learning of graph neural networks: a unified review (2022). ArXiv:2102.10757 [cs]
12. Kirchhoff, Y., et al.: Skeleton recall loss for connectivity conserving and resource efficient segmentation of thin tubular structures (2024)

An Approach Based on Networks and Machine Learning for Gastric Cancer Treatment Recommendation

Lucas Queiroz Melo da Costa, Carlos Henrique Costa Ribeiro, and Emmanuel Dias-Neto

Abstract Cancer is a health challenge for modern societies, as it affects millions of people every year and brings a heavy burden in terms of treatment costs, the population's quality of life and survivability. For oncologists, the treatment approach is also surrounded by uncertainties due to the unknown phenomena related to patients' response, and precision medicine plays a significant role in defining the best therapeutics. This paper proposes to leverage network science and machine learning techniques in an integrated approach to provide a wide spectrum of information—from qualitative networked analysis to treatment response predictions—in order to assist practitioners' decision for a neoadjuvant-based scheme when treating gastric cancer patients. As a caveat, the framework also tackles the challenges of implementation in a real case scenario by using data from the oncology clinical practice in A.C. Camargo Cancer Center, a leading institution for cancer treatment in Brazil. Put in context, this work is part of an undergoing project towards a comprehensive AI-based decision support system for neoadjuvant treatment applied to gastric cancer.

Keywords Networks · Machine learning · Cancer treatment

L. Q. M. da Costa (✉)
Federal University of São Paulo, São José dos Campos, SP, Brazil
e-mail: lqm.costa@unifesp.com

C. H. C. Ribeiro
Aeronautics Institute of Technology, São José dos Campos, SP, Brazil
e-mail: carlos@ita.br

E. Dias-Neto
A. C. Camargo Cancer Center, São Paulo, SP, Brazil

M. Macedo et al. (eds.), *Complex Networks XVI*, Springer Proceedings in Complexity,
https://doi.org/10.1007/978-3-031-93619-7_5

1 Introduction

1.1 Background and Motivation

Cancer constitutes a major challenge to human health worldwide, a disease characterized by an uncontrolled cell growth caused by gene mutations which can affect a multitude of tissues and organs, hence manifesting through many subtypes and distinct prognosis [10]. Gastric cancer (GC) is one of the most common and lethal cancer manifestations, almost reaching 14,000 deaths in Brazil in 2020 while presenting an estimate of 21,000 occurrences per year within the country between 2023–2025. Its causes are yet not fully understood, mainly carried by Helicobacter pylori infection but also related to individuals' genetic background, excessive ingestion of industrialized food, behavior and even occupation patterns [5].

Not only in its manifestations, cancer treatment is also surrounded by unknown phenomena [2]. Currently, a standard approach in the oncology practice for GC involves the use of chemotherapy right after diagnosis (neoadjuvant phase), followed by a surgery that usually implies in the removal of most or all the stomach, and a continuation of the chemotherapy to remove residual disease (adjuvant phase). This approach is called "perioperative" and has historically demonstrated to be the most effective life-saving strategy when treating GC, leading to a greater overall survival of patients. On the other hand, it is observed that a very significant share of the afflicted individuals does not show a clear favorable response, as these subjects present stable or even disease progression during the preoperative period. In this context, it becomes crucial to correctly identify and predict which patients would benefit from the neoadjuvant chemotherapy, supporting the decision-making process of the practitioner when recommending a treatment scheme for GC. This constitutes an important step towards a more effective treatment and survivability, while preserving the non-responders that would need to receive alternative drugs or go straight to surgery.

Artificial Intelligence (AI) and Network Science (NS)—particularly Complex Networks and Machine Learning—have gained prominence as a great ally not only in cancer diagnostics (Al-Azzam and Shatnawi 2021) but also in predicting the success of treatment approaches and assessing survivability of patients [12]. Through clinical and multi-omics data, computerized tools can learn from medical history by a combination of compositional and relational information to anticipate if an individual is most likely to benefit or not from a treatment choice, hence structuring the problem at hand as a clusterization analysis and prediction task. Dealing with medical data, however, is not simple [7]. Real, new and reliable data is hard to find, as proved by the recurring need to emulate medical information or "recycle" old, publicly available and ubiquitous datasets to develop new research. When new streams of "real-world" data are available to be collected, it is usually hard to gather it in a structured and consistent way [13], and there are persistent systematic phenomena related to critical diseases and treatments' information which require tailored analysis [19].

Considering such challenge, this work aims to study the phenomena related to the neoadjuvant treatment for gastric cancer patients under the view of clinical oncology data, hence implementing a networked, machine learning-integrated approach that proposes to enhance practitioners' analysis capabilities and substantiate the decision regarding the recommendation of such treatment scheme.

In order to develop a purposeful study and maximize real-world impacts, this work is conducted within the context of the partnership between the Aeronautics Institute of Technology (ITA) and A.C. Camargo Cancer Center (ACC) to obtain the best quality data, medical and technical expertise. The methodology outlined in this work represents part of a work yet in progress, as a promising steppingstone towards a comprehensive decision support toolkit for practitioners' assistance when dealing with gastric cancer patients and the neoadjuvant treatment scheme. The framework presented is based on a preliminary network analysis from [3] and guided under the operating conditions of ACC's clinical oncology practice.

1.2 Literature Review

As a preliminary step towards the development of any computational tool, data processing is key to guarantee quality and consistency to the results yet to be obtained. In this subject, data-centric AI rises as a hot topic, understanding that curating and working the data is not a merely operational process—as a "stock" step to be input to learning tasks—but a critical phase to promote reliability and accuracy of models to its real-world applications. Contrary to the "model-centric AI" view, a data-centric lens seeks to systematically characterize, evaluate and improve the underlying data to train and evaluate models [15], understanding algorithmic refinement as a less important—and in some cases even solved—problem. Whang et al. [17] addresses a data-centric view based on the premises that quality data is paramount to machine learning, even for the state-of-the-art techniques. Looking at the oncology and clinical field, Adeoye et al. [1] admits that the concept of data-centric AI is still incipient in healthcare systems. As main gaps, the authors point to data imbalance and fairness affecting data quality and hence limiting discriminatory performance in structured datasets.

Looking from a networked perspective over the medical problem at hand, some key developments are highlighted to provide the foundation for Complex Networks' modelling. A keystone article is that of Ma and Zhang [8], where a powerful procedure for CN generation is proposed when facing the problem of combining multi-omics data to identify patients from different cancer subtypes. The first procedure is set to generate CNs for each data source, subsequently fusing these multiple networks by an original framework of "Affinity Network Fusion" (ANF). Experiments to enhance the zero-shot clustering performance are carried and results point that better performance can be reached by just using two out of the three omic graphs, but integrating all three omics provide more consistent classification among all cancer subtypes.

Diving into the proposal of this work, a landmark constatation through the literature reviewed is regarding the many insightful developments found when exploring the interconnections between Complex Networks and Machine Learning. Torshizi and Petzold [16] develop a semi-supervised ML model based on graphs which are composed of many different genomic data in the form of "biological pathways" to classify ovarian cancer. Three different types of genomics are collected and translated into networks by a K-Nearest Neighbors approach. At the same time, the most helpful genes within each biological pathway are selected to form complementary graphs considering three different approaches for generation. Graph learning models are trained on these structures and results surpass many other state-of-the-art techniques. Looking at different health applications for CNs in combination with ML, Renjini et al. [14] modelled Complex Networks by using a correlation map extracted from cough sounds. Many time series data points are grouped in order to form nodes, and correlations between those are capped to form edges. Network metrics such as Average Path Length (APL), Graph Density and Degree Centrality of nodes are used as input to ML methods, which are capable of classifying sound between cough, croup and pertussis.

Lastly, a key insight when approaching the problem through Machine Learning applications is related to data processing challenges, which are well spotted in literature specially related to medical and oncology tasks. Kotsiantis et al. [6] propose a review on techniques to handle imbalanced datasets, stressing the fact these occurrences can arise naturally in many fields and that traditional classifiers are not well suited for this context. They highlight techniques to address imbalanced ML data such as under/oversampling, feature selection and one-class learning. AUC are among the recommended metrics to assess such imbalanced learning problems.

2 Methodology and Modelling

Inspired by the reviewed approaches that relate to the challenges envisaged for this work, a phased approach is proposed to fulfill the objectives of this paper, segmented between gathering and processing datasets, carrying out Complex Networks analysis and implementing Machine Learning classification tasks, which are described below.

2.1 Datasets

The data utilized for this work is composed from two different sets, called "Neoadjuvacy" (NEO) and "Multiple Myeloma" (MMRF).

Neoadjuvancy (NEO): Related to the neoadjuvant treatment and extracted from AC Camargo (ACC) restricted data source, with guidance and expertise from its Medical Genomics Group to only select relevant clinical attributes for the prediction task on the response to the neoadjuvant scheme for gastric cancer treatment. For preprocessing, some criteria were imposed to guarantee information quality. For the targeted feature for analysis, the attribute "percentage of viable cells in the surgical specimen" (free translation of the Portuguese "porcentagem de células viáveis na peça cirúrgica", related to the surgical specimen and the primary tumor) was recommended by ACC experts as the marker for treatment success, with a suggested threshold of 10% as the limit below which the patient can be considered a "good respondent" to the neoadjuvant scheme. Such target choosing and preprocessing led to a total of 28 remaining features and 66 (out of the total 265) unlabeled instances. The binarization procedure resulted in a class imbalance of 35 (17.6%) "good" and 164 (82.4%) "bad" responders.

Multiple Myeloma (MMRF): As a comparative exercise, a public dataset regarding Multiple Myeloma (MM), provided by the National Cancer Institution [11] is utilized. To keep a coherent comparison with the NEO case, only clinical attributes from the MMRF dataset are considered, totalizing 59 features. The feature "disease_status" was chosen as target for prediction as it states the prognostic response to the MM, and a binarization of its categories led to a near-perfect balance of 550 (49.5%) "survivors" and 560 (50.5%) "deceased" after treatment, with no unlabeled instances.

Common to both Complex Networks and Machine Learning phases, the remaining missing values are tackled by imputation using the respective attribute's mode and average, in the case of categorical and numerical features, respectively. Data normalization is also conducted in its MinMax approach. For the Complex Networks, unlabeled instances are removed, and class imbalances are allowed to occur. The Machine Learning analysis handles imbalances by implementing the ADASYN data augmentation [4], and the curse of dimensionality is mitigated by implementing feature extraction through the FAMD method [9]. The MMRF set has the same processing pipeline applied as the NEO data but does not require imputation/augmentation as it does not present missing values/imbalanced class ratios.

2.2 Complex Networks

The Complex Network set of analysis aims at providing qualitative intelligence over the relationship structure among patients, as well as extracting relational information to be integrated into the Machine Learning classification tasks with the objective of enhancing its predictive performance. The proposed methodology is segmented between stages of network modelling, topology analysis and feature extraction.

Network modelling. When laying out the networks, patients are set as nodes and their relationships are translated by connecting links. A main idea here explored relates to the development of two different generation pipelines by varying the distance/similarity calculation and the rationale for link formation.

Regarding the distance/similarity calculation, two methods are considered:

Minkowski Distance (MD). Calculates the Minkowski metric for distance between objects (nodes) of a given dataset, following (1).

$$dist(x_i, x_j) = \sqrt[p]{\sum_{l=1}^{d} |x_{il} - x_{jl}|^p} \quad (1)$$

where $p = 1$ for features with boolean values (Hamming distance) and $p = 2$ for rational-valued attributes (Euclidean distance), being l from 1 up to d as the number of features in the dataset and x_i, x_j two given nodes (objects) in the network.

Hybridized distance (HD). Inspired by the methodology of Zhang and Ma (2018), calculates the pair-wise Euclidean Distance matrix among objects and a local diameter vector considering each node's average distance to its "k" closest neighbors, where "k" is a pre-defined parameter. Then, a weighted average distance measure (balanced by a pre-defined "α" parameter) is calculated by combining the Euclidean distance between each pair of nodes and their respective node diameters. Lastly, a Gaussian kernel is applied to minimize noisy signals and convert the distance measure to similarity values, which are then compiled for each pair of objects within the network.

Regarding the rationale for link formation, two approaches are developed:

Network Distance Threshold (NDT). Considers applying a threshold over the entire pair-wise distance matrix previously calculated, as a percentile of the lowest distances found for the entire network. In this sense, only pair-wise distances below such threshold are linked.

Node-specific Similarity Threshold (NsST). Sets quantile values to select only the pre-defined highest share of distances/similarities for each node's specific profile, which are then connected in the structure.

Topology analysis. Based on the generated "Natural" networks, an analysis over the topology obtained is proposed to gather qualitative insights. More specifically, an analysis is carried out by segmenting the class-specific relational behavior of features, to better understand the part played by each attribute to group and provide insights regarding patient's prognostics.

Feature extraction. Based on the generated "Subnetworks", information specific to each of the patients within each of the subnetwork is extracted. The usability of such information can be structured as new features to be inserted in a Machine Learning classification task and calculated in the occasion of inserting a new patient in the network itself, providing additional parameters to assist practitioners' analysis. Specifically, class-related values from its "k" neighbors are analyzed. And an exhaustive search procedure is implemented to determine an optimal "k" to be considered in each case.

2.3 Machine Learning

Two Machine Learning classification tasks are implemented with regards to the datasets NEO and MMRF, aiming at predicting each set's targeted feature. The tasks are named as "Standard Learning" and "Developed Learning", described below.

Standard Learning. To set the baseline for what would be the learning performance achieved by the "standard" approach to classification tasks, a learning round is implemented using classical techniques. Such diversity of methods is proposed under the interpretation of the *No Free Lunch* theorem [18] for the Machine Learning context. The data processing applied covers only the main steps used for the Complex Network analysis. A leave-one-out learning loop is implemented and metrics captured in the form of AUC and Recall—which constitute key indicators of learning performance for the case on small, imbalanced datasets.

Developed Learning. This learning procedure is proposed in a way to address some of the data issues spotted during previous Standard Learning and Complex Network analysis. The datasets are segmented according to the "subset" feature groupings and the features extracted from the respective "subnetworks" are incorporated into the sets. Inspired by the reviewed literature for this work, an extended set of classifiers are implemented, now also covering algorithms based on gradients, graph-based and one-class learning techniques. Internal parameters are also optimized along an internal cross-validation routine within the leave-one-out learning loop. Data processing steps are executed as for the Standard Learning, with additional data augmentation and feature extraction procedures implemented for the NEO set through ADASYN [4] and FAMD [9] techniques, respectively.

3 Results and Discussion

3.1 Complex Networks

Through the forementioned methodology, networks with patients as nodes and class labels (treatments response or outcome) are modelled and presented in Fig. 1, for datasets NEO ('a', 'c') and MMRF ('b'). Two generating pipelines are established by matching *MD* calculation/*NDT* linkage methods and *HD* calculation/*NsST* linkage procedures, which originate the so-called "Natural" ('a', 'b') and "Subnetworks" ('c') structures. The latter is obtained by a data-centric idea of segmenting the datasets' features into subsets, hence providing a dedicated view for the different dimensions of the medical problem. For the NEO set, subgroupings are established by segregating features related to patients' "Clinical Profile (PC)", "Disease Manifestation" (MD) and "Treatment Parameters" (PT) information. For the MMRF set, subgroupings are established by features segmented among "Clinical Profile (PC)", "Exam" (EX) and "Treatment" (TR) dimensions.

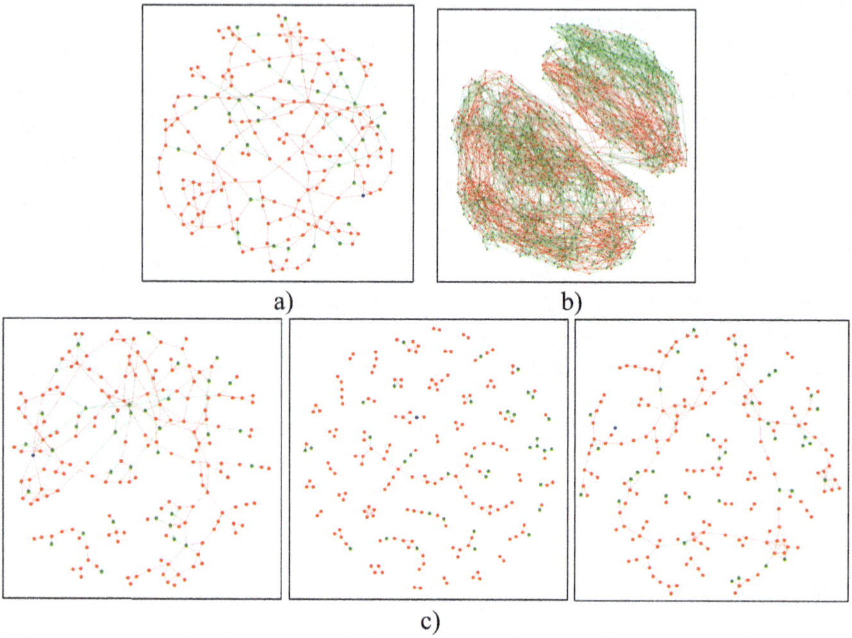

Fig. 1 Natural networks for dataset **a** NEO and **b** MMRF, alongside subnetworks for dataset **c** NEO and subnetworks PC (left), MD (center) and PT (right)

Starting at the NEO natural network ('a'), a first impression relates to the reduced number of nodes due to the lower number of objects present in this dataset. In addition, it is to note the relative lower density of links formed between nodes if compared to other sets, which is intentional of the NDT threshold applied as to highlight the main relationships between nodes while keeping a connected structure. Regarding the spatial distribution, local groupings with a denser concentration of links are observed, connected among each other by somewhat linear "strings" of nodes. This represents an indicative of patients with stronger internal similarity forming clusters of nodes, which in turn are inter-related by short and linear structures of connections. When adding label (color) information to the analysis, it becomes evident the imbalanced nature of the data, with low count of "good" responders to the neoadjuvant treatment scheme (green color). This minority class is not clearly segregated within the network, being scattered amid the majority "bad" responders (red color). Such mixing of different response levels to treatment increases the challenges of clearly distinguishing patients by their prognostics, which is useful in oncology practice as to analyze a new patient by his/her neighborhood when inserted in the network. Nevertheless, it is possible to observe some closeness between the green nodes, in many cases positioned as low-order neighbors between themselves.

The natural MMRF network ('b') obtained from the MMRF presents two clearly distinct groupings connected by only a few nodes, which prompts for further investigation over the characteristics of such regions and the characteristics of the "connector" patients. Adding label information, for the major grouping located on the top right area of the visualization it is possible to identify a cluster with a predominant class of "good" outcome to the cancer treatment (green color), alongside a mixed cluster with "bad" outcome to cancer treatment as a predominant class (red color). This translates to a more "predictable" outcome for the MM treatment based on the relationships of a potential new patient with the historical afflicted individuals' data. Switching to the subnetworks for the NEO set ('c'), an observation is the fragmented aspect of the nodes' linkage patterns when considering the different subsets of features separately in each network, resulting in multiple chains of few connected objects. Nevertheless, the structures modelled do point to a more identifiable relationship among the "good" responders (green nodes), as most of them appear to be connected to another similarly labeled object at least in the second order. This key insight is critical for the latter phases of analysis of this work, where each subnetwork in Fig. 1 ('c') will have object-specific features extracted and inserted in the respective subset's machine learning tasks.

Continuing the topology characterization of the networks based at datasets' features and their part at bringing together objects from the same class, the diagram in Fig. 2 presents the average distance between nodes (patients) from the same class (treatment response) when considering each feature in isolation for the NEO data.

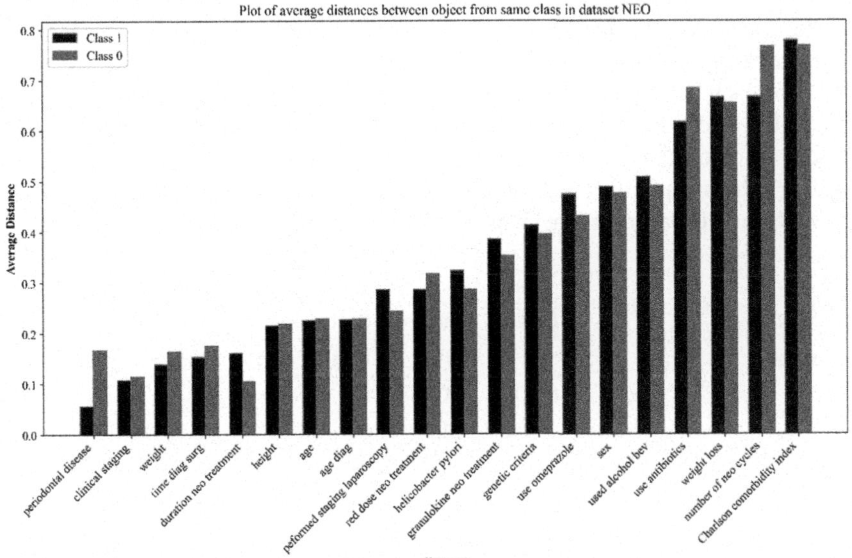

Fig. 2 Average distance between objects calculated by each feature, for each class in natural network NEO

From the diagram, it is observable that there are features which act better as a 'node magnet' for a given class, while keeping the opposite label more disperse. That is the exemplary case of attributes 'periodontal disease' and 'duration net treatment' for classes of 'good' (1) and 'bad' (0) responders, respectively. There are also variables that promote a lower or higher overall distance between nodes for both classes, as exemplified by 'clinical staging' and 'Charlson comorbidity index' which appear to equally attract and repulse same-class objects for both labels, respectively. In any case, these types of attributes could serve as preliminary potential markers for patients' clustering, whether focusing to group objects from a specific class ('periodontal disease' and 'duration net treatment') or both labels simultaneously ('clinical staging').

Finally, when working on the forementioned idea of extracting features from the networked structures, the average response of the "*k*" closest neighbors for nodes within a subnetwork is analyzed. This signal becomes useful when studying the yet unknown potential response to the neoadjuvant treatment for a given "new patient" that is held under scrutiny, while also aggregating useful relational information regarding each of the dataset's objects which can enhance the Machine Learning prediction capabilities. In this sense, Table 1 summarizes results obtained for the gap analysis among the targeted feature values of the '*k*' closest neighbors for each node in the network, while also conducting an extensive search to find the best '*k*' parameter for each subset and the entire dataset, considering both NEO and MMRF data.

It is clarified that the label values are presented in range from 0 to 100 for the NEO set, while assuming discrete values from 0 to 4 for the MMRF case. In an overall sense, it is noticeable that all average label gaps found do not surpass the absolute value of 50% of the range between all possible labels, which would undermine the value of such relational information—if true, each node could assume any value for its targeted feature from the standpoint of its closest neighbors. For the NEO case, it is to note the rather broad variation of the optimal '*k*' found for each subset, with 'PT' being optimized at a much lower number of closest neighbors considered. Nevertheless, the average label gap is kept rather similar across all subsets' analysis

Table 1 Best "K" neighbors' analysis to extract label information

Dataset	Subset	Characteristic	
		Best "k" neighbors	Average label gap
NEO	PC	12	35.171
	MD	15	34.641
	PT	4	34.489
	All	12	34.974
MMRF	PC	14	1.249
	EX	25	1.199
	TR	1	1.027
	All	26	1.173

in a range close to 35, which means that the optimal number of closest neighbors from each patient (node), on average, have a variation of 35 in the values of their targeted attribute that conveys the treatment response level for the neoadjuvant approach (in a scale from 0 to 100). The MMRF presents an average gap around the marks of 1.1 (approximately 39% of the label range), with label information from 1 up to 25 neighbors optimally considered for each node under different subnetworks.

3.2 Machine Learning

Finally, as a Machine Learning exercise for the NEO and MMRF classification problems, Table 2 presents the learning metrics obtained by the baseline "Standard" and tailored "Developed" approaches, the latter including relational information from the previously modelled subnetworks in the form of new features, which in turn are extracted for each patient based on the optimal "K" neighbors parameters found for datasets NEO and MMRF during previous analysis. For note, the classifiers implemented are abbreviated as K-nearest neighbors (KNN), Random Forests (RF), Support Vector Machine (SVM), Logistic Regression (LR), XGBoost (XGB), Isolation Forest (IF) and Label Propagation (LP). The subsets for the NEO data are taken as.

From the Recall metric calculated for the "Standard Learning" case, it is noted that while the MMRF set presents reasonable values (from 0.76 to 0.83), the NEO data results in near-zero figures. The latter outcome indicates that the model was unable to predict a single true positive object for the Neoadjuvant task, hence "reason guessing" all instances as negative-labeled under the influence of the imbalanced nature of the dataset. The AUC metrics for the NEO case are set with an average of 0.469, below the 0.5 threshold which is considered as the "random classification" mark (such mark could be obtained by just randomly selecting a class for each patient). The MMRF dataset, however, presents consistent AUC values with an average of 0.947, which is attributed by the increase data size and more equal proportion of objects belonging to each class.

Due to the inherent inability of the classical learners to predict the Neoadjuvant treatment prognostics, the tailored "Developed Learning" approach is applied to the NEO set exclusively, which brings into account the network-extracted features to increase the amount of information leveraged for the prediction task, implements data augmentation to mitigate the imbalance issue and add more learners relying in different rationales which could prove more useful for originally-imbalanced and reduced datasets. As a result, the average AUC is raised by 15% and surpasses the "random" 0.5 threshold, reaching an average of 0.539 and even isolated marks of 0.707 for the combination of subset "Treatment Parameters" and learner "Logistic Regression" (LR). This implies a critical increase in classification quality, in which the machine learning architecture goes from erroneous contribution to positive impact over the prognostics' prediction. Furthermore, although not rising significantly, the Recall metric does not reach the almost-zero figures anymore, which translates that

Table 2 Machine learning tasks for datasets NEO and MMRF, considering multiple learners and subsets (1) clinical profile, (2) disease manifestation and (3) treatment parameters

Approach	Standard				Developed	
Dataset	NEO		MMRF		NEO	
Method/metric	Recall	AUC	Recall	AUC	Recall	AUC
KNN	0.029	0.553	0.760	0.868	0.257 (1) 0.400 (2) 0.457 (3)	0.458 (1) 0.490 (2) 0.579 (3)
RF	0.000	0.554	0.816	0.976	0.200 (1) 0.229 (2) 0.229 (3)	0.499 (1) 0.496 (2) 0.564 (3)
SVM	0.000	0.228	0.791	0.964	0.400 (1) 0.486 (2) 0.657 (3)	0.470 (1) 0.576 (2) 0.614 (3)
LR	0.057	0.543	0.827	0.979	0.486 (1) 0.429 (2) 0.571 (3)	0.573 (1) 0.532 (2) 0.707 (3)
XGB	–	–	–	–	0.143 (1) 0.314 (2) 0.314 (3)	0.599 (1) 0.529 (2) 0.558 (3)
IF	–	–	–	–	0.114 (1) 0.229 (2) 0.057 (3)	0.411 (1) 0.543 (2) 0.526 (3)
LP	–	–	–	–	0.486 (1) 0.026 (2) 0.486 (3)	0.563 (1) 0.484 (2) 0.550 (3)

the learners are now not biased by the imbalanced nature of the NEO data and hence are operating under a "fair" effort to actually predict the treatment's outcomes for gastric cancer patients.

4 Conclusions

This work presented a Network and Machine Learning approach to tackle the issue of treatment recommendation for gastric cancer. Based on the real-world dilemma of recommending a neoadjuvant treatment scheme for patients, data from the oncology practice of the A.C. Camargo Cancer Center was utilized (NEO) alongside a comparative publicly available (MMRF) set to generate qualitative insights and quantitative metrics to support the practitioners' decision-making, enhancing their analytic capabilities to deal with high volumes of data which in turn present intricate relationship dynamics.

At first, Complex Networks were generated under different methodologies to multiply the relational information available. Visual inspection of the so-called "Natural Networks" revealed challenges to group same-class patients in the NEO task, while the MMRF presented clearer distinction patterns. For the NEO data, an analysis of features and their capabilities of grouping same-class objects was implemented and revealed potential markers for patients' clustering. Looking to provide new features for the Machine Learning task based on the relationship between nodes, an extensive search across different number of neighbors to be accounted was executed and the optimal value was utilized, minimizing the targeted feature gap between each patient and its considered neighborhood. Finally, the Machine Learning task under the baseline "Standard" approach—which does not consider measures for data augmentation nor the network-extracted features—revealed difficulties in handling the imbalanced data, which was not observed in the MMRF set. As a countermeasure, the implementation of the "Developed" approach considering procedures of data augmentation, network-extracted features and additional learning methods was able to increase the learning capabilities and bring more confidence over the predictions' outputs.

As future directions of research, the multitude of relational and predictive information provided by the Networks and Machine Learning tasks can be further studied to provide even more insights and quantitative metrics to substantiate the cancer treatment recommendation, while also being able to be framed as a comprehensive decision support tool for the oncology practice. More specifically, the networks generated can introduce new patients under scrutiny to be analyzed in terms of their location and relationship dynamics among other historical patients. On the Machine Learning side, the multiple predictions based on subsets can be further consolidated by the usage of meta-learning approaches, and the incorporation of network-based information into the training rounds can "customize" the learning process for each new patient analyzed.

Acknowledgements This study was partially financed by the "Coordenação de Aperfeiçoamento de Pessoal de Nível Superior—Brasil" (CAPES). This work was also carried out with the support from the São Paulo Research Foundation (FAPESP), Brazil, grant n. 2014/26897-0 (project: "Epidemiology and genomics of gastric adenocarcinomas in Brazil"). The research uses data from A.C. Camargo Cancer Center's Oncology practice, authorized by the Ethics and Research Committee (registration n. 2134/15).

References

1. Adeoye, J., Hui, L., Su, Y.-X.: Data-centric artificial intelligence in oncology: a systematic review assessing data quality in machine learning models for head and neck cancer. J. Big Data **10**, 28 (2023)
2. Coelho, F., Braga, A.P., Natowicz, R., Rouzier, R.: Semi-supervised model applied to the prediction of the response to preoperative chemotherapy for breast cancer. Soft. Comput. **15**, 1137–1144 (2011)

3. Costa, L.Q.M., Ribeiro, C.H.C., Dias-Neto, E.: A study of networks for decision support in neoadjuvant treatment for gastric cancer. In: Galoá Proceedings. LVI Brazilian Operations Research Symposium, SBPO (2024)
4. Imbalanced learn user guide for over-sampling. https://imbalanced-learn.org/stable/over_sampling.html. Last accessed 15 Nov 2024
5. INCA Síntese de Resultados e Comentários. https://www.gov.br/inca/pt-br/assuntos/cancer/numeros/estimativa/sintese-de-resultados-e-comentarios. Last accessed 15 Nov 2024
6. Kotsiantis, S., Kanellopoulos, D., Pintela, P.: Handling imbalanced datasets: a review. GESTS Int. Trans. Comput. Sci. Eng. **30**, 25–36 (2006)
7. Li, J., Tian, Y., Li, R., Zhou, T., Li, J., Ding, K., Li, J.: Improving prediction for medical institution with limited patient data: leveraging hospital-specific data based on multicenter collaborative research network. Artif. Intell. Med. **113**, 102024 (2021)
8. Ma, T., Zhang, A.: Affinity network fusion and semi-supervised learning for cancer patient clustering. Methods **145**, 16–24 (2018)
9. Max Halford—Prince. https://maxhalford.github.io/prince/. Last accessed 15 Nov 2024
10. Murthy, N.S., Bethala, C.: Review paper on research direction towards cancer prediction and prognosis using machine learning and deep learning models. J. Ambient Intell. Humaniz. Comput. **14**, 5595–5613 (2021)
11. National Cancer Institute Genomic Data Commons: The Multiple Myeloma Research Foundation. https://gdc.cancer.gov/about-gdc/contributed-genomic-data-cancer-research/foundation-medicine/multiple-myeloma-research-foundation-mmrf. Last accessed 15 Nov 2024
12. Povoa, L.V., Ribeiro, C.H.C., Silva, I.T.: Machine learning predicts treatment sensitivity in multiple myeloma based on molecular and clinical information coupled with drug response. PLoS One, (7), e0254596 (2021)
13. Povoa, L.V., Calvi, U.C.B., Lorena, A.C., Ribeiro, C.H.C., Silva, I.T.: A multi-learning training approach for distinguishing low and high risk cancer patients. IEEE Access **9**, 115453–115465 (2021)
14. Renjini, A., Swapna, M.S., Raj, V., Kumar, K.S.: Complex network-based pertussis and cough analysis: a machine learning approach. Phys. D **433**, 133184 (2022)
15. Seedat, N., Van der Schaar, M.: Data-Centric AI, https://www.vanderschaar-lab.com/data-centric-ai/. Last accessed 15 Nov 2024
16. Torshizi, A.D., Petzold, L.R.: Graph-based semi-supervised learning with genomic data integration using condition-responsive genes applied to phenotype classification. J. Am. Med. Inf. Assoc. **25**(1), 99–108 (2018)
17. Whang, S.E., Roh, Y., Song, H., Lee, J.-G.: Data collection and quality challenges in deep learning: a data-centric AI perspective. VLDB J. **32**, 791–813 (2023)
18. Wolpert, D.H., Macready, W.G.: No free lunch theorems for optimization. IEEE Trans. Evol. Comput. **1**(1), 67–82 (1997)
19. Zhu, Y., Zhu, X., Kim, M., Yan, J., Kaufer, D., Wu, G.: Dynamic hyper-graph inference framework for computer assisted diagnosis of neurodegenerative diseases. IEEE Trans. Med. Imaging **38**(2), 608–615 (2019)

Population Dynamics in the Global Coral-Symbiont Network Under Temperature Variations

Maria Gabriella Cavalcante Basílio and Daniel Ratton Figueiredo

Abstract Coral reefs are crucial to marine biodiversity and rely on a delicate symbiotic relationship between corals and zooxanthellae algae. Water temperature variations, however, disrupt this association, leading to coral bleaching events that severely affect marine ecosystems. This study presents a mathematical model for the population dynamics of coral and symbiont species considering the coral-symbiont network and recurrent warming events. The model incorporates thermal tolerances of species and coupled growth dynamics (between corals and symbionts) to investigate how network structure and thermal tolerance influence the species' growth. Using real data from different ocean regions, results reveal that network connectivity plays a significant role in population growth after successive warming events, with generalist species demonstrating greater growth across all regions analyzed. The comparatively higher correlation between node degree and final population also emphasizes the impact of ecological network structure on species growth, offering valuable insights into coral reef population dynamics under climate change. This research highlights the need to consider network structure beyond species' thermal tolerances when evaluating the ecological responses of corals to environmental changes.

Keywords Population dynamics · Ecological networks · Coral bleaching

1 Introduction

The symbiotic relationship between species of coral and microalgae, known as zooxanthellae, is fundamental to the survival of coral reefs, as algae and corals exchange nutrients with each other. However, when these organisms are exposed to warming events, the bond between the host coral and endosymbiotic algae breaks down. The

M. G. C. Basílio (✉) · D. R. Figueiredo
Department of Computer Science and Systems Engineering (PESC), Federal University of Rio de Janeiro (UFRJ), Rio de Janeiro, Brazil
e-mail: basiliog@cos.ufrj.br

D. R. Figueiredo
e-mail: daniel@cos.ufrj.br

M. Macedo et al. (eds.), *Complex Networks XVI*, Springer Proceedings in Complexity,
https://doi.org/10.1007/978-3-031-93619-7_6

breaking of this bond is called coral bleaching [1, 8] because the color of corals often depends on the type of algae associated with them. Therefore, when the association is broken, the coral loses its color and becomes white, a phenomenon known as bleaching.

The bleaching event causes several ecological impacts such as a decrease in its growth rate. Furthermore, bleaching events are recurrent and have become more frequent in the last decade [4]. However, warming events do not equally affect all organisms, as they have different thermal tolerances which represents their capacity to grow under water temperature variation [8, 9]. Finally, the capacity to grow also depends on the symbiotic relationships of the organism. Intuitively a coral with multiple and diverse algae has a higher tolerance to warming events and thus more robust growth.

This work presents a mathematical model for the growth of corals and algae considering both their symbiotic relationship and thermal resistance under recurrent warming events. In particular, a bipartite network encodes the symbiotic relationships and a differential equation for each organism (coral and algae) captures its population growth. This equation depends on the network structure, thermal tolerances, water temperature and the population of the symbionts.

Using real data collected from different ocean regions, the proposed growth model is applied to a representative model for recurrent warming events. Starting from identical initial populations, the model shows that different organisms have very different growth patterns over time. The relationship between population and network structure and thermal tolerance is investigated. Results indicate that, for different ocean regions, correlation between network structure and population is stronger than thermal tolerance and population. This highlights the importance of the symbiotic network on understanding bleaching events.

2 Related Work

Understanding the consequences of rising ocean temperatures in the development of coral reefs through network analysis has been broadly explored in the recent literature. A notable work studies the global network between coral species and *Symbiodiniaceae* and its resistance to temperature stress as well as its robustness to temperature perturbations [8]. Another recent work proposes and evaluates an eco-evolutionary model that shows that shortcuts in the dispersal network (e.g., corals that disperse larvae throughout the ocean to coral reefs) across environmental gradients (i.e., changes in non-living factors through space or time) hinder the persistence of population growth across regions [5, 6]. Theses works have been quite successful in identifying how the network structure affects the sensitivity of corals to changes in water temperature, either in symbiotic associations networks or in dispersal networks.

For instance, in the context of the global coral-symbiont network [8], null networks were created by altering physiological parameters of organisms or the network structures. A bleaching model was developed with weighted links representing temperature thresholds for host-symbiont pairs. Resistance to temperature stress and ecological robustness were assessed by analyzing how different networks responded to increasing temperatures (e.g., link removal) and species (e.g., node) removal. Results indicated that robustness to bleaching and other perturbations varied across spatial scales and differed from null networks. The global coral-symbiont network was more sensitive to environmental attacks, such as rising temperatures, with symbionts providing more stability than hosts. Network structure and thermal tolerances are not represented by uniform random patterns, making the system more vulnerable to environmental changes.

The dispersal networks represent demographic connectivity between populations located in different habitats. These networks describe how offspring of species move between these habitats, forming connections that influence both the demography and the growth of populations [5, 6]. Additionally, through the eco-evolutionary model, it was observed that random networks performed better in non-evolving populations, while regular networks favored populations with higher evolutionary potential [6]. These networks, by reducing maladaptive gene flow, allowed local populations to adapt more efficiently. Results reinforce the importance of considering eco-evolutionary dynamics, network structures, and environmental gradients when assessing species' ability to migrate and persist under climate change.

3 Data Source and Network

Data from the GeoSymbio [2] and a complementary database [8] were used to construct the bipartite coral-symbiont network. Geosymbio database provided information about the organisms, such as *Symbiodinium* type based on ITS2 sequence type, scientific name (genus and species) of coral and the location (i.e., ocean region) from which the *Symbiodinium* specimen was collected. There is a total of 53 ocean regions in the GeoSymbio. The complementary database was used to obtain data on the thermal tolerances of the *Symbiodinium* type and the coral host. Unfortunately, the database is not complete and some organism do not have a specified thermal tolerance. In such cases, the mean value of the thermal tolerance was used as reference.

A bipartite network encoding the relationship between host corals and its endosymbiotic algae was generated for each ocean region. In particular, an edge represents the symbiotic relationship between a symbiont species and a host species in the ocean region where it were observed (Fig. 1). Thus, there are no edges between organisms of distinct regions. Besides, each node represents a symbiont species or host species in a region. For instance, if a same species of symbiont or host occurs in k regions, then the network will have k vertices of this species.

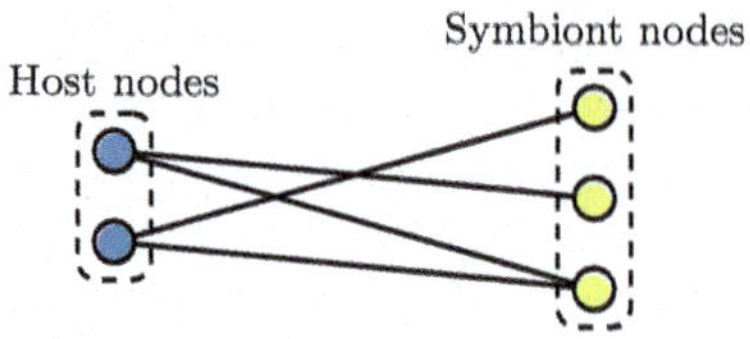

Fig. 1 Bipartite coral-symbiont network with host nodes (blue) and symbiont nodes (yellow)

Table 1 Network nodes and edges in different connected components analyzed

Region	Symbiont nodes	Host nodes	Edges	Density
Great Barrier Reef	76	198	415	0.055
Phuket	36	152	442	0.162
Western Indian	43	131	337	0.120
Western Caribbean	36	61	111	0.101
Florida	26	32	75	0.180

Table 2 Degree in the global coral-symbiont network

Type of node	Standard deviation	Minimum degree	Average degree	Maximum degree
Symbiont	8.306	1	3.168	102
Host	2.346	1	2.332	51

The global coral-symbiont network has 867 symbiont nodes and 1178 host nodes, 2747 edges and 181 connected components. Note that the connected components are at least the number of ocean regions (i.e., 53), however the global network has many more connected components. Hence, there are multiple connected components within the same ocean region.

Moreover, five connected components of the global network each corresponding to a different region were chosen to be analyzed separately, as shown in Table 1. These regions represent the most threatened regions of coral bleaching in the oceans. Note that these networks have different number of nodes and edges, but relatively similar edge density.

3.1 Degree Distribution

The degree of the global coral-symbiont network is analyzed, considering all 53 ocean regions. Table 2 shows that the average degree of both types of nodes is relatively similar but not the standard deviation which is larger for the symbiont nodes.

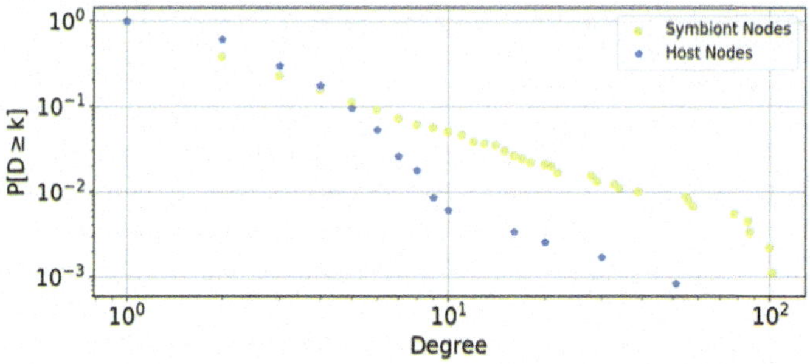

Fig. 2 The complementary cumulative distribution function (CCDF) of symbiont and host nodes

Furthermore, since the minimum degree is 1, there are no isolated organisms (nodes) in the coral-symbiont network.

Figure 2 shows the complementary cumulative distribution function (CCDF) for the degree of both symbiont and host nodes. Note that both distributions are heavy-tailed since a tiny number of symbionts are connected 100 or more hosts and a tiny number of hosts are connected to 50 or more symbionts. Further, note that the symbionts have a heavier tail (the distribution curve decreases more slowly) indicating that symbionts connect more, also because the number of host nodes is much larger.

Moreover, the difference between the tail values and the average degree values of the two types of nodes is very significant. Recall that the average degree of the symbionts and hosts are approximately 3 and 2, respectively. Thus, the majority of hosts and symbionts are specialists (have very few connections) while a tiny amount of both nodes are generalists (have large number of connections).

4 Mathematical Model for Population Growth

A novel population model using the network, thermal resistance, and symbiotic population was developed with the aim of studying the population dynamics of coral and algae under exposure to recurrent warming events. In what follows the model is described in detail and Table 3 presents variables and parameters of the model.

In essence, the model is a system of coupled ordinary differential equations to track the population of symbionts and hosts over time. This model considers the coral-symbiont network, where every node has associated with it a population. Note that this model uses a single variable per node instead of a variable for each symbiotic relationship (i.e., edges). Consequently, this model has significantly fewer variables (see Table 1). However, network edges drive the population dynamics as growth of corals and algae are coupled and symbiotic.

Table 3 Definition for symbols of variables and parameters of the model

Symbol	Definition (variables and parameters)
$S_i(t)$	Population of the ith symbiont species at time t
$H_i(t)$	Population of the ith host species at time t
N_i^s	Neighborhood of the ith symbiont species
N_i^h	Neighborhood of the ith host species
r_i^s	Population growth rate of the ith symbiont species
r_i^h	Population growth rate of the ith host species
m_i^s	Population mortality rate of the ith symbiont species
m_i^h	Population mortality rate of the ith host species
τ_i^s	Thermal tolerance of the ith symbiont species
τ_i^h	Thermal tolerance of the ith host species

Let $S_i(t)$ and $H_i(t)$ denote the population of symbiont i and host i at time t, respectively. The evolution (derivative) of S_i over time is given by:

$$\frac{dS_i}{dt} = \frac{S_i}{|N_i^s|} r_i^s \left(\sum_{j \in |N_i^s|} \frac{H_j}{|N_j^h|} \right) - S_i m_i^s \qquad (1)$$

Note that there is a growth term (positive) and a mortality term (negative) that are driven by a growth rate (r_i^s) and mortality rate (m_i^s). Moreover, the growth term also depends on the network. This is the main contribution of the proposed model. In particular, the growth rate depends on the population of the corals that have a symbiotic relationship (edge) with this symbiont.

In particular, the growth rate is multiplied by the sum across the neighboring hosts of the fraction of the host populations (H_j) divided by its neighbors (N_j^h). This fraction is assumed to interact with a fraction of this symbiont population, which is given by S_i divided by its neighbors (N_i^s). Thus, for each neighboring host j, the growth rate is multiplied by $\frac{H_j}{N_j^h} \frac{S_i}{N_i^s}$. Note that the second term does not depend on j.

Dividing the population of an organism by its degree assumes that each population interacts uniformly with the population of neighboring organism. This normalization ensures that the interaction of a symbiont or host population is distributed evenly among its connections. While hosts typically have fewer neighbors than symbionts, this asymmetry is inherent to the network structure and is represented in the model by this normalization. Moreover, this assumption significantly simplifies the model as it requires a single variable (population) for each node while also capturing network heterogeneity (different degrees).

Table 4 Parameter definitions and values used in simulations

Parameter	Value	Definition
r_0^s	1.0	Scaling factor for symbionts' growth rate [6]
r_0^h	1.0	Scaling factor for hosts' growth rate [6]
z	29.1 °C	Optimum growth temperature for symbionts and coral hosts
μ	0.3	The base mortality [7]

The growth rate (r_i^s) is given by:

$$r_i^s = \frac{r_0^s}{\sqrt{2\pi\left(\tau_i^s\right)^2}}.e^{\left(\frac{-(T(t)-z)^2}{\left(\tau_i^s\right)^2}\right)} \tag{2}$$

While the mortality rate (m_i^s) is given by:

$$m_i^s = \begin{cases} \mu, & \text{if } \; T(t) \leq z \\ 1 - e^{\left(\frac{-(T(t)-z)^2}{\left(\tau_i^s\right)^2}\right)}, & \text{if } \; T(t) > z \end{cases} \tag{3}$$

Note that both the growth and mortality rates have already been proposed in the literature [6, 7] and depend on the current local sea temperature ($T(t)$) and thermal tolerance (τ_i^s) of each organism.

Considering the mortality rate, note that if the temperature is lower than or equal to the optimum temperature for growth (given in the model by parameter z), the mortality rate is equal to μ (see value in Table 4). However, if the current local sea temperature is higher than the ideal growth temperature, the mortality rate is a function that depends on the thermal tolerance of the organisms.

The evolution (derivative) of H_i over time is given by:

$$\frac{dH_i}{dt} = \frac{H_i}{|N_i^h|} r_i^h \left(\sum_{j \in N_i^h} \frac{S_j}{|N_j^s|} \right) - H_i m_i^h \tag{4}$$

Note that this equation is identical to (1) making the model symmetric. The growth rate and mortality rate for hosts are also given by (2) and (3), respectively (replacing superscript s with h, as shown in Table 3). Thus, there is no inherent population growth advantage between symbionts and hosts. Of course, their growth depends on the parameters of the model such as network structure, thermal tolerance, water temperature and initial population.

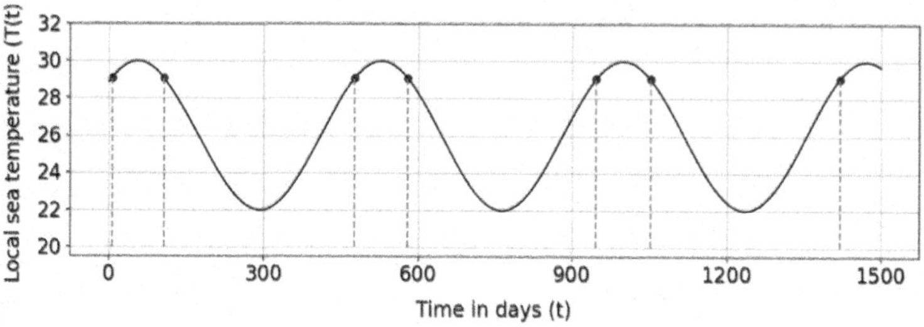

Fig. 3 Local sea temperature function over time. The blue dashed lines represents the moment at which the model reached the optimal growth temperature (z)

The growth and mortality rate of symbionts and hosts depend on the current local sea temperature. Thus, a model for the evolution of the sea temperature is needed. The temperature model used in this work was based on real ocean temperature data, collected over 26 months, in two regions of Western Australia: Coral Bay and Tantabiddi [3] which is shown to be recurrent. In particular, the temperature model is given by:

$$T(t) = 4\cos\left(\frac{t}{75} + 30.6\right) + 26 \tag{5}$$

The choice of parameters for the temperature model was arbitrary to emulate recurrence within a temperature range and timescale. Figure 3 shows the evolution of the temperature over time indicating the optimal growth temperature value.

Finally, (1) and (4) will be solved numerically and independently for each region (see Table 1) according to the above temperature model over a time horizon that simulates successive warming events over 1500 d, as shown in Fig. 3.

An extended mathematical model based on the above can be found in the ArXiv.

5 Quantitative Analysis

Numerical solution of the population model provides insights into how populations of host corals and endosymbiotic algae behave when exposed to successive warming events when their growth is coupled by the network. Moreover, assuming that all host and symbiont species have some initial population, it is possible to characterize the role of the symbiotic interactions network structure in the growth dynamics of these populations and how the network influences recovery after warming events.

In particular, all symbiont species have an initial population of 1000, while all host species have an initial population of 100. Thus, there is no preferred species at time zero.

5.1 Population Growth

Figure 4 shows the population growth for symbionts and hosts at the Great Barrier Reef region. Note that all species showed an overall increasing trend in the population.

Moreover, when water temperature is far from the optimal the population of most species decreases. This same trend was observed in all other regions. However, all species showed resilience as they continue to grow in population despite the thermal stress events.

Nevertheless, even with the initial populations being the same for all species, differences in the evolution of populations occur due to the structure of the network and thermal resistances. Since the network structure is not uniform, as the node degrees are very different, population growth is also not uniform. Figure 5 shows the population distribution after 1500 d for both symbionts and hosts for all regions analyzed. Note that population distribution for hosts exhibits a heavy tail in all regions.This highlights the central role of network structure in population dynamics, as it determines how the species interactions shape resilience. Species with higher node degree (generalist species) adapt better to environmental changes, while species with lower node degree (specialist species) grow slower when exposed to thermal disturbances.

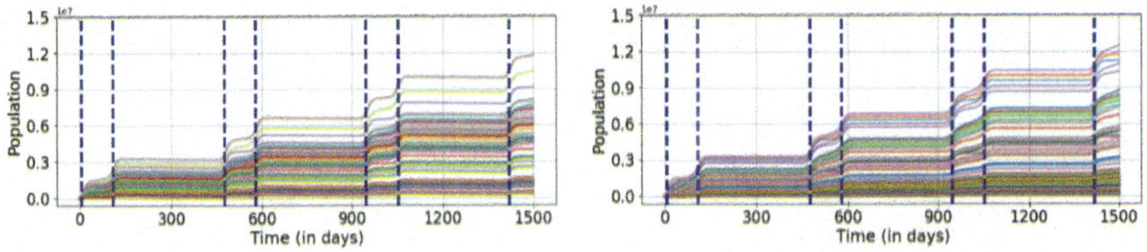

(a) Great Barrier Reef - Symbionts' growth (b) Great Barrier Reef - Hosts' growth

Fig. 4 Population dynamics of symbiont species (**a**) and host species (**b**) at Great Barrier Reef over time. Blue dashed lines indicate the moment when the optimum temperature (z) for growth was reached

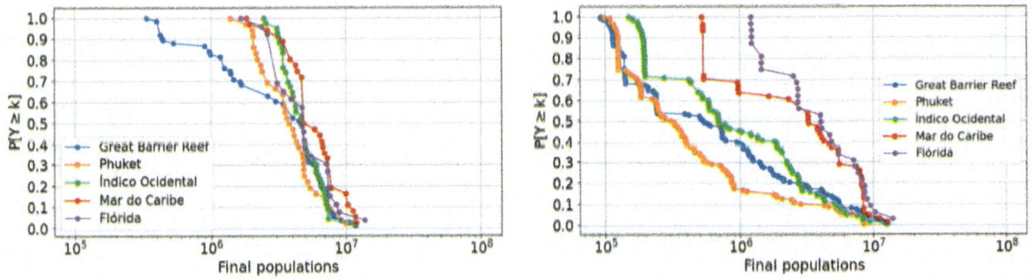

(a) CCDF of symbionts' final populations (b) CCDF of hosts' final populations

Fig. 5 CCDFs of symbionts' (left) and hosts' (right) final populations in their respective collection region

5.2 *Influence of Network Structure*

In order to study whether the structure of interactions (network) influences the resistance of species to thermal stresses, a random bipartite network was created for each region (as shown in Table 1) using a previously described methodology [8]. In particular, each original edge was repositioned uniformly at random, destroying any biological symbiotic affinity. Note that the number of edges of each network was preserved. Moreover, the network randomization procedure adopted did not allow any isolated nodes, as all nodes in the randomized network have degree of at least 1. Figure 6 shows the population distribution for each region when growing on the random networks. Interestingly, the distributions have a much lighter tail in comparison to the original networks (see Fig. 5).

The role of the network structure and the thermal tolerances of species on the population can be studied through correlation analysis.

Table 5 shows the correlation between the thermal tolerances of species and their final populations. Note that this value is relatively low across all regions (close to zero). Therefore, final population are not even moderately correlated with the thermal tolerance.

In contrast, Table 6 shows the correlation between the final population and node degree. Note that this correlation is relatively larger than with thermal tolerance for all ocean regions. Moreover, for three regions the correlation is above 0.3 (considered

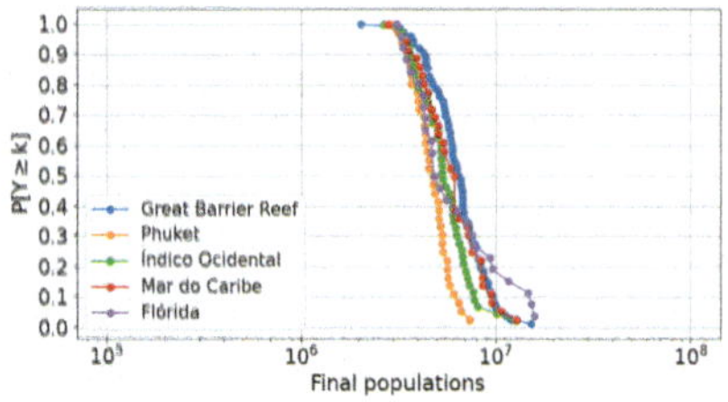

(a) CCDF of symbionts' final populations

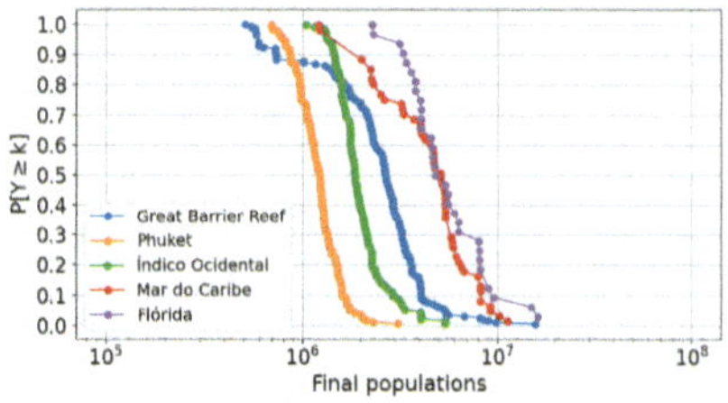

(b) CCDF of hosts' final populations

Fig. 6 CCDFs of symbionts' (left) and hosts' (right) final populations in their respective collection region in the random networks

Table 5 Correlation between final populations and thermal tolerances

Region	Symbiont nodes	Host nodes
Great Barrier Reef	0.134	−0.085
Phuket	−0.008	−0.022
Western Indian	0.133	−0.130
Western Caribbean	0.205	0.132
Florida	0.263	−0.151

Table 6 Correlation between final populations and node degrees. For the random network, the sample average correlation and its standard deviation is reported using 50 independent instances of the random network.

Region	Real network		Random network	
	Symbionts	Hosts	Symbionts	Hosts
Great Barrier Reef	0.384	0.552	0.213 ± 0.072	0.168 ± 0.073
Phuket	0.193	0.186	0.232 ± 0.144	0.078 ± 0.087
Western Indian	0.183	0.206	0.075 ± 0.164	0.100 ± 0.092
Western Caribbean	0.469	0.579	0.154 ± 0.125	0.146 ± 0.134
Florida	0.383	0.404	0.078 ± 0.216	0.084 ± 0.202

Table 7 Correlation between final populations and sum of degrees of neighbors

Region	Symbiont nodes	Host nodes
Great Barrier Reef	0.181	−0.235
Phuket	0.132	−0.365
Western Indian	0.127	−0.499
Western Caribbean	0.308	−0.229
Florida	0.105	−0.238

a moderate value) for both symbionts and hosts. In biological terms, this indicates that symbionts and corals that are generalists (have higher degree) are able to growth faster than those that are specialists (have lower degree) in the presence of water temperature variation.

Table 6 also shows the correlation between degree and final population in the random networks. Note that the average correlation for the random networks is considerably smaller than the original networks that have moderate correlation (above 0.3). As expected, randomly repositioning the edges removes the heavy tail property and makes the degree distribution more centered. These results reinforce the importance of network structure for the survival of these organisms.

Finally, Table 7 shows the correlation between the sum of degrees of neighbors and the final populations of each species. Differently from degree correlation, negative correlations values for the hosts stand out. Note that, in all analyzed networks, host nodes are more numerous (Table 1). Therefore, symbiont nodes have many more neighbors than host nodes (Fig. 2). Thus, in (4), the growth of the hosts, determined by r_i^h and network structure, is harmed by the large number of neighbors of the symbionts, since the sum of fractions are smaller due to their larger denominator ($|N_j^s|$). Hence, the lower the degree of the hosts' neighbors, the larger the sum fractions (4) in the contribution to the growth rate of the hosts. This relationship is an indicative of the negative correlation between these two variables.

On the other hand, this correlation for symbionts is always positive, although weak, since their neighbors tend to have smaller degrees which will increases their growth rate, determined by r_i^s and network structure. Thus, correlation of neighbors degrees is not symmetric between hosts and symbionts, differently from degree (where three regions had moderate correlation for both hosts and symbionts).

An extended numerical evaluation of the extended model can be found in the ArXiv

6 Conclusion

This paper investigated the population dynamics within the global coral-symbiont network under temperature variations, with a focus on the impact of thermal stress on coral bleaching. Using a bipartite network model, the relationships between coral hosts and their symbiotic algae has been characterized, identifying how network structure influences population growth and resilience to recurrent warming events. Besides the numerical analysis, a main contribution of this work is a simple and parameterized mathematical model capturing the network structure.

Our results demonstrated that the network structure plays a crucial role in determining the capacity of coral and symbiont species to recover from warming events, with generalist species exhibiting stronger recovery patterns.

Furthermore, correlations between final populations and node degrees emphasized the importance of network connectivity in population growth. These findings enhance the understanding of the ecological factors that contribute to coral reef resilience and underscore the need to consider network structure when evaluating species adaptability to climate change.

Acknowledgements This research received partial funding from grants by the following Brazilian agencies CNPq, FAPERJ and CAPES.

References

1. Donner, S.D., Skirving, W.J., Little, C.M., Oppenheimer, M., Hoegh-Guldberg, O.: Global assessment of coral bleaching and required rates of adaptation under climate change. Glob. Change Biol. **11**(12), 2251–2265 (2005)
2. Franklin, E.C., Stat, M., Pochon, X., Putnam, H.M., Gates, R.D.: Geosymbio: a hybrid, cloud-based web application of global geospatial bioinformatics and ecoinformatics for symbiodinium–host symbioses. Mol. Ecol. Resour. **12**(2), 369–373 (2012)
3. Fulton, C.J., Depczynski, M., Holmes, T.H., Noble, M.M., Radford, B., Wernberg, T., Wilson, S.K.: Sea temperature shapes seasonal fluctuations in seaweed biomass within the Ningaloo coral reef ecosystem. Limnol. Oceanogr. **59**(1), 156–166 (2014)
4. Hughes, T.P., Kerry, J.T., Baird, A.H., Connolly, S.R., Dietzel, A., Mark Eakin, C., Heron, S.F., Hoey, A.S., Hoogenboom, M.O., Liu, G., et al.: Global warming transforms coral reef assemblages. Nature **556**(7702), 492–496 (2018)

5. McManus, L.C., Forrest, D.L., Tekwa, E.W., Schindler, D.E., Colton, M.A., Webster, M.M., Essington, T.E., Palumbi, S.R., Mumby, P.J., Pinsky, M.L.: Evolution and connectivity influence the persistence and recovery of coral reefs under climate change in the Caribbean, Southwest Pacific, and coral triangle. Glob. Change Biol. **27**(18), 4307–4321 (2021)
6. McManus, L.C., Tekwa, E.W., Schindler, D.E., Walsworth, T.E., Colton, M.A., Webster, M.M., Essington, T.E., Forrest, D.L., Palumbi, S.R., Mumby, P.J., et al.: Evolution reverses the effect of network structure on metapopulation persistence. Ecology **102**(7), e03381 (2021)
7. Walsworth, T.E., Schindler, D.E., Colton, M.A., Webster, M.S., Palumbi, S.R., Mumby, P.J., Essington, T.E., Pinsky, M.L.: Management for network diversity speeds evolutionary adaptation to climate change. Nat. Clim. Change **9**(8), 632–636 (2019)
8. Williams, S.D., Patterson, M.R.: Resistance and robustness of the global coral–symbiont network. Ecology **101**(5) (2020)
9. Williams, S.D.: Corals are more than the sum of their colonies: a network science perspective on the role of coral complexity and its consequences for coral reef health. Ph.D. thesis, Northeastern University (2020)

Dengue Serotypes Cyclicity Evidenced by the Impact-Frequency Histogram of the Visibility Graph

L. L. Lima and A. P. F. Atman

Abstract Epidemics are one of the most significant challenges throughout history, and climate change has contributed to the increasing frequency and diversity of outbreaks. It is the case of dengue fever, which has been responsible for hundreds of thousands of cases worldwide in the last decades. The proliferation of different serotypes and their association with other diseases have made the situation too complex, and there is an urgent demand for alternative approaches to fully understanding the dynamics of the outbreaks. In this work, we apply the visibility graph approach to analyze the time series of dengue occurrence patterns in two large Brazilian cities. We introduce a new impact-frequency histogram protocol to evaluate cyclic dengue patterns in a single plot. We analyzed the time series of cases and estimated a period for re-infestation of the disease. The tool has proven helpful in analyzing temporal series, especially for epidemic diseases.

Keywords Network analysis · Vector-borne diseases · Epidemics

L. L. Lima (✉)
Norwegian Research Center—NORCE, Norce, Norway
e-mail: larissalopeslima@yahoo.com.br

Programa de Pós-Graduação em Modelagem Matemática e Computacional—PPGMMC/CEFET-MG, Belo Horizonte, Brazil

A. P. F. Atman
Departamento de Física, Centro Federal de Educação Tecnológica de Minas Gerais—CEFET-MG, Belo Horizonte, Brazil
e-mail: atman@cefetmg.br

Programa de Pós-Graduação em Modelagem Matemática e Computacional—PPGMMC/CEFET-MG, Belo Horizonte, Brazil

Instituto Nacional de Sistemas Complexos—INCT-SC/CEFET-MG, Belo Horizonte, Minas Gerais, Brazil

M. Macedo et al. (eds.), *Complex Networks XVI*, Springer Proceedings in Complexity,
https://doi.org/10.1007/978-3-031-93619-7_7

1 Introduction

The dynamics of contemporary society are complex and subject to unforeseen factors, such as climate change. The emergence of epidemics and the expansion of endemic regions are examples of the challenges facing humanity nowadays, with disastrous consequences for society and the economy, such as the recent coronavirus crisis. A remarkable example of adaptation to this complex scenario is the possible expansion of infectious diseases to other areas due to climate change [1].

Dengue fever, for instance, is a vector-borne disease transmitted mainly by the *Aedes aegypti* mosquito. Established more than three centuries ago [2, 3], this disease wiped out thousands of people and is a threat present in tropical countries, such as Brazil, where millions of people are in risk areas. This country recorded several epidemics of dengue from the second half of the 19th century, going through a period of epidemiological silence and reappearing in 1986 when it became endemic and a national public health problem [4, 5]. Dengue is a cyclical disease with major outbreaks every three to five years [6].

Dengue dynamics is a complex problem dominated by environmental determinants of transmission [7]. The incidence of this disease is associated with the combination of rainy seasons, high temperatures, winds, and elevation [8]. Moreover, the urban environment introduces heterogeneity in the reproduction sites of the *Ae. aegypti* mosquito, influencing the disease transmission dynamics through this vector [9]. Besides environmental aspects, high-density cities with massive population mobility and low socioeconomic status are a perfect scenario for disseminating the disease. There may be other factors that make the dynamics of dengue epidemics even more complex, such as those indicated by some studies pointing to the existence of immunological interactions between serotypes [10, 11]. The arising of new serotypes is currently a primary difficulty in fully understanding the dengue dynamics, playing a key role in disseminating the disease between regions [12].

Dengue forecasting research has increased in the last few years [13], but there is still a long way to go due to all the complex factors involved in this disease. Knowing the disease dynamics contributes to corroborating or reorienting surveillance and control actions, so it is possible to optimize resources to control the disease [3] and better understand how it spreads among the population. Thus, in this paper, we propose a new protocol to assist the public agents and the research community in quantifying and analyzing the dengue dynamics. This protocol allows them to estimate how severe the epidemics are and offers a way of analyzing dengue outbreaks.

We applied the Visibility Graph (VG) technique to analyze epidemic time series data, which maps the time series into a graph. Besides, we introduce the impact-frequency histogram to analyze the recurrence time of dengue cycles. It was noted that more than a simple statistical analysis is needed to address the complexity of the problem under study by treating the data. In the Visibility Graph technique, the graph reflects several properties arising from the data series used in its construction, and its application can help interpret information that is not easily visualized in the original series [14]. Therefore, this study aims to obtain the characteristics of the disease (such as time to recurrence) through the analysis of the visibility graph generated from the time series. As dengue recurrence is complex, the VG technique is critical to analyzing long-range correlations.

2 Methods

The proposed use of the Visibility Graph and Horizontal Visibility Graph (HVG) techniques arose due to our need to validate a computational model developed to simulate dengue spreading. Initially, we aimed to validate it by comparing the recurrence time of disease outbreaks. However, we could not find a way to compare the data using infection peak counts or comparing frequencies. By performing this analysis, it was essential to define the size of the peak that would be considered an outbreak or not, which may cause a bias for the analysis. This bias is eliminated using VG and HVG techniques since tiny peaks typically have fewer connections than larger peaks (larger peaks play a role as barriers to connecting smaller peaks and others). The proposed protocol is described in this section.

2.1 Dengue Data

This work used 12-year data set from dengue cases recorded in two Brazilian cities (Fig. 1): Rio de Janeiro (epidemiological week number 34, 2001 to epidemiological week number 3, 2014) and Belo Horizonte (epidemiological week number 1, 2007 to epidemiological week number 16, 2019), both from the Brazilian Notifiable Diseases Information System (*Sistema de Informação de Agravos de Notificação* [15]—SINAN, in Portuguese) data set. These data include only confirmed dengue cases per epidemiological week.

Though there are different time intervals, it is worth noticing that the distribution of disease cases over time is different for each city. It happens because the two cities have different dynamics and differences in population characteristics, weather conditions, and other factors that interfere with disease dynamics [12].

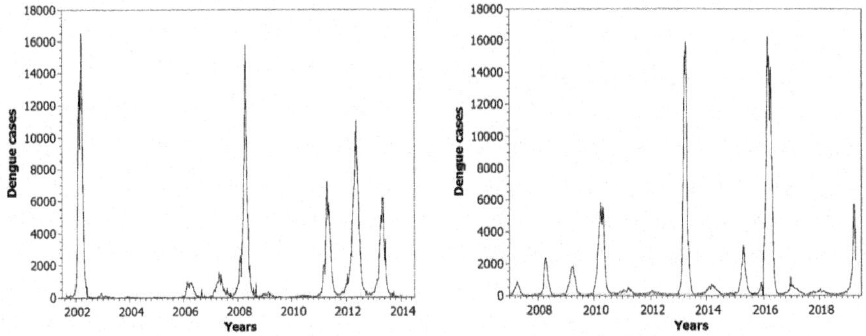

(a) Rio de Janeiro (b) Belo Horizonte

Fig. 1 Number of dengue cases registered over 12 years in Rio de Janeiro (RJ) and Belo Horizonte (BH) (SINAN [15] data)

Belo Horizonte (Minas Gerais state) has an area of 331.354 km^2 and a population of 2315560 people (2022) [16]. The city has a tropical climate, with an average monthly rainfall of 276 mm from November to March and 41 mm from April to October, and an average annual temperature of 21.1 °C [17]. On the other hand, Rio de Janeiro (state of Rio de Janeiro) has an area of 1200.329 km^2 and a population of 6211223 habitants (2022) [16]. The climate in Rio de Janeiro is also tropical, with an average annual temperature of 23.2 °C [18].

2.2 *Visibility Graph*

The Visibility Graph technique consists of building a graph from a time series following the procedure proposed by Lacasa et al. [14]. Each point is a node; two nodes are connected if they meet the visibility criteria. It consists of drawing a line between two data points and verifying if this line does not cross any other data. If so, the two nodes are connected.

The mathematical formulation of the VG is the following: two arbitrary values (t_a, y_a) and (t_b, y_b) have visibility (and are connected) if any other value (t_c, y_c) located between them fulfills (1) (considering $t_a < t_c < t_b$):

$$y_c < y_b + (y_a - y_b)\frac{t_b - t_c}{t_b - t_a}. \tag{1}$$

The Horizontal Visibility Graph [14] was also used in our analysis. In the HVG, two nodes are connected if we can draw a horizontal line joining y_a and y_b without crossing any intermediate data y_c (considering $t_a < t_c < t_b$). In this work, both VG and HVG were applied only to nonzero data points, i.e., the link between two nodes is only possible if both are greater than zero ($y(t) > 0$). Negative numbers were not considered since the time series are from dengue cases.

2.3 *Impact-Frequency Histogram*

Here, we introduce the Impact-Frequency histogram (IFH), a tool to be used along with the VG technique that quantifies the frequency of the epidemic outbreaks mediated by their impact (the number of infected individuals).

The IFH is built considering each link obtained in the VG of the time series of the number of dengue cases. The impact of a given link is calculated by summing the number of occurrences in each node divided by the difference between them (a unity is added in the denominator to avoid division by zero). Taking two arbitrary values (t_a, y_a) and (t_b, y_b), the impact γ is given by:

$$\gamma(\Delta t) = \frac{y_a + y_b}{|y_a - y_b| + 1}, \quad (2)$$

where $\Delta t = |t_a - t_b|$ is the corresponding period (inverse of the frequency).

3 Results and Discussion

A significant concern in epidemiological studies of dengue fever is to determine if there is a pattern for the outbreak's occurrence [19]. As shown in Fig. 1, although an outbreak is expected yearly, a precise determination of the period is not straightforward since it depends on several factors such as population dynamics and environmental conditions [12]. To overcome these limiting factors to the analysis, we apply the VG technique and build the IFH for two large Brazilian cities with quite different characteristics, such as climate, elevation, and population dynamics.

Figures 2 and 3 show the VG and HVG built from SINAN data recording of Rio de Janeiro and Belo Horizonte, respectively. The VG plot is shown in Figs. 2a and 3a, and it is possible to notice some hierarchical structures. Smaller hubs connect to larger ones, each connected to local nodes. Also, the VG figures show closely connected central nodes. These nodes are fundamental for connecting other parts of the graph, representing the biggest peaks of dengue outbreaks.

In the HVG figures (Figs. 2b and 3b), the topology is presented more clearly, highlighting the hierarchical organization. The network appears to be more evenly distributed in Belo Horizonte (Fig. 3b) compared to Rio de Janeiro (Fig. 2b). Additionally, a comparison between VG and HVG reveals that all HVG links are included within the VG link set. This suggests that HVG functions as the backbone of VG in this analysis. Moreover, HVG may provide a more representative depiction of this study, as VG tends to overemphasize the influence of links between data from the same dengue outbreak-a limitation mitigated by HVG.

It is also possible to compare the network morphology of multiple time series, even if they originate from different periods. In the case of the VG, the graph for Rio de Janeiro exhibits a highly connected, hierarchical structure, characterized by one dominant hub alongside smaller hubs. In contrast, the graph for Belo Horizonte features hubs of more comparable sizes. This comparison highlights that the sensitivity of VG and HVG techniques in capturing features within time series surpasses what can be identified through descriptive statistics alone, such as the average number of peaks.

Two significant infection peaks were recorded in Rio de Janeiro: the first occurred during epidemiological weeks 1–22 in 2002, and the second between epidemiological weeks 1–26 in 2008 (Fig. 1a). The second is a more central peak in the time series and prevents the first half of the time series from connecting to the other in the VG. However, this central peak connects to both halves of the graph, forming the most prominent hub in Fig. 2a. A similar pattern is observed in the HVG (Fig. 2b), where the central peaks interrupt direct connections between the left and right sides of the

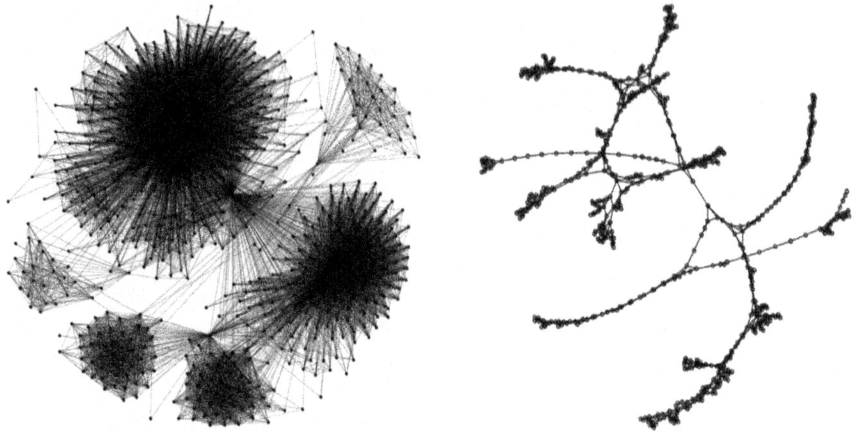

(a) Visibility Graph (b) Horizontal Visibility Graph

Fig. 2 Visibility graph of the number of cases in Rio de Janeiro from 2001 to 2014

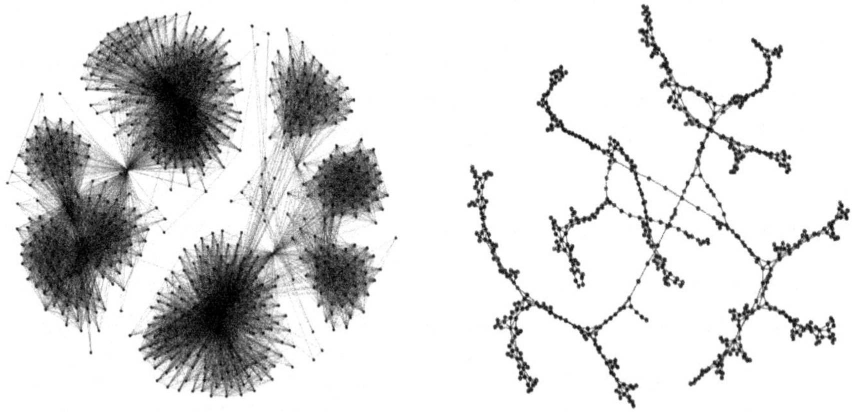

(a) Visibility Graph (b) Horizontal Visibility Graph

Fig. 3 Visibility graph of the number of cases in Belo Horizonte from 2007 to 2019

graph. Additionally, a third peak in the second half of the series further divides the data, hindering connections between these sections of the graph.

In Belo Horizonte, the highest peaks were observed during weeks 3–28 in 2013 and weeks 1–26 in 2016 (Fig. 1b). Like the Rio de Janeiro graph, the most central peak in Belo Horizonte's data divides the graph, preventing connections between the first and second halves. Additionally, a prominent peak in the second half further segments the graph, which may explain why the hubs are smaller and exhibit greater homogeneity compared to the Rio de Janeiro graph (Fig. 2a), as shown in Fig. 3a. In the initial weeks, smaller peaks are observed, and while the VG can establish connections between these early peaks and the central peak, these connections are not detected by the HVG. This is due to the fourth peak (between weeks 1 and 39 in 2010) blocking horizontal connections between them.

3.1 Impact-Frequency Histogram

The analysis of the Impact-Frequency Histogram graphs for the VG and HVG in Rio de Janeiro (Fig. 4) and Belo Horizonte (Fig. 5) reveals a significant difference between the two municipalities. For both techniques, the distribution of distances between peaks ranging from 0–10 to 0–20 weeks can be disregarded, as these peaks are associated with the same outbreak.

A statistical test was conducted to assess the normality of the data. Given the large dataset, the Kolmogorov-Smirnov test was used (R version 3.6.1 [20]). The results indicated that the impact-frequency data do not follow a normal distribution ($\alpha = 0.05$). Consequently, the non-parametric Kruskal-Wallis test was applied for comparison, revealing a significant difference between the data ($\alpha = 0.05$).

The IFH results reveal that the VG captures periodicity over longer time ranges, while the HVG is more sensitive to short-term cyclicity. This difference arises from the HVG's reliance on horizontally connected peaks, resulting in fewer connections due to its shading effect. The HVG is particularly susceptible to disruptions caused by prominent infection peaks, where a large peak can hinder connections between years that are otherwise distant.

When analyzed as a periodic function, the impact-frequency results display distinct patterns for Rio de Janeiro and Belo Horizonte, reflecting differences in disease spread behaviors in the two municipalities. In Rio de Janeiro, the VG (Fig. 4a) shows high impact values for periods between 35 and 55 weeks, corresponding to the annual recurrence of dengue epidemics. Additional high-impact values occur over well-defined intervals: 90–110 weeks (approximately two years), 140–160 weeks (2.7–3 years), 190–210 weeks (3.6–4 years), 250–260 weeks (4.8–5 years), and 290–315 weeks (5.6–6 years). The HVG (Fig. 4b) similarly identifies the annual infection peak (35–55 weeks). Notably, the HVG also highlights high-impact values at the same intervals as the VG, but with a significantly smaller number of points.

By comparing the HVG and VG data with the number of infected cases in Rio de Janeiro, it becomes evident that the annual recurrence of dengue infections was most frequent in 2006–2008 and, to a lesser extent, in 2009. Additional outbreaks were observed in 2011–2014. The periods corresponding to the observed impact values align with the infection data as follows: two years (2006–2008), 2.7–3 years (2008–2011), 3.6–4 years (2008–2012), and 5.6–6 years (2002–2008).

For the HVG, annual connections were identified between 2006 and 2007, 2007 and 2008, 2011 and 2012, 2012 and 2013, and 2013 and 2014. Other periods, though less represented, include two years (2011–2013), 2.7–3 years (2008–2011), 3.6–4 years (2008–2012), and 5.6–6 years (2002–2008).

In Belo Horizonte, the VG results (Fig. 5a) similarly reflect the annual recurrence of dengue observed in Rio de Janeiro. Additionally, two distinct periods show

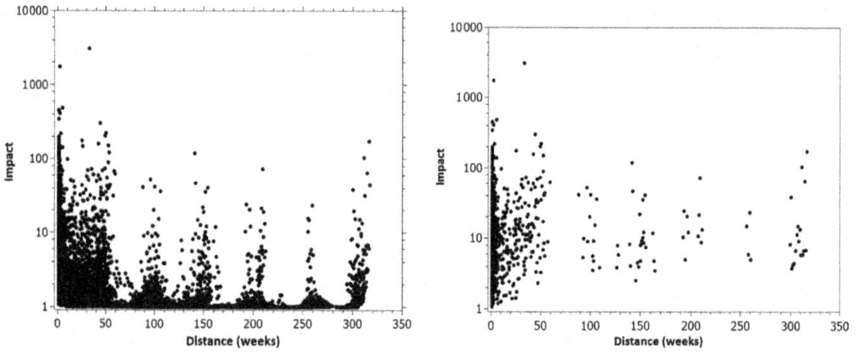

(a) Visibility Graph (b) Horizontal Visibility Graph

Fig. 4 Impact as a function of distance (period) for the Visibility graph and the Horizontal Visibility graph of SINAN [15] data for Rio de Janeiro

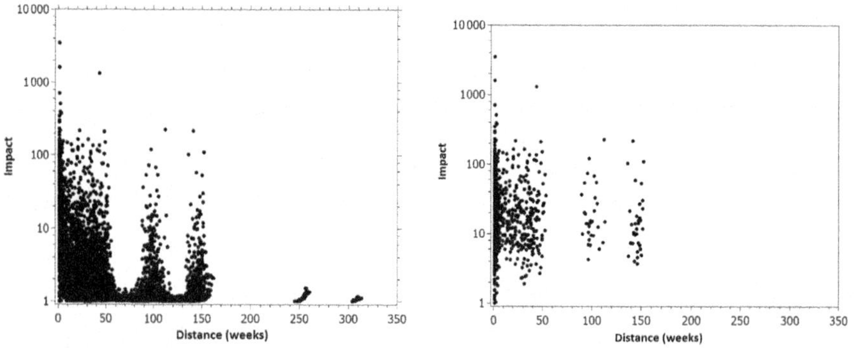

(a) Visibility Graph (b) Horizontal Visibility Graph

Fig. 5 Impact as a function of distance (period) for Visibility graph and Horizontal Visibility graph of SINAN [15] data for Belo Horizonte

high impact values: 85–115 weeks (1.6–2.2 years) and 135–160 weeks (2.6–3 years). Some impact values were recorded around 250 weeks (4.8 years) and 300–315 weeks (5.75–6 years). The HVG results (Fig. 5b) exhibit a similar pattern, with notable impacts at approximately 50, 100, and 150 weeks.

Based on weeks and the number of infected individuals, the comparison of impact data in Belo Horizonte indicates that annual infections are evident almost yearly, though weaker in some instances, such as in 2012 and 2018. For other years, the impact periods align with infection years, similar to the pattern observed in Rio de Janeiro. These include 1.6–2.2 years (2008–2010, 2011–2013, 2013–2015, 2016–2018, 2017–2019), 2.6–3 years (2007–2010, 2010–2013, 2013–2016, 2016–2019), 4.8 years (between certain weeks from 2008 to 2013), and 5.75–6 years (between certain weeks from 2007 to 2013). In addition to the annual infection cycles observed with the VG, the HVG results include similar periods of 1.6–2.2 years (2008–2010 and 2013–2015) and 2.6–3 years (2010–2013, 2013–2016, and 2016–2019).

3.2 *Limitations and Strengths*

Climate changes as heatwaves increase the relative risk of dengue fever [21], and other studies try to find a relationship between this disease and biological and environmental factors that characterize the recurrence patterns [17, 22]. Our study tested a new protocol that showed that the association of the visibility graph technique and the impact-frequency histogram was able to quantify the impact of the different peaks of dengue over the years, characterizing the occurrence of dengue in two Brazilian cities.

However, the IFH is a new technique that needs more exploration. One of the biggest limitations is quantifying occurrences of the same outbreak, that is, dengue cases from the same peak being counted in the analysis. In this case, the HVG showed a cleaner result, but a specific study is still required to determine how this limitation interferes with the results. We suggest a specific study focusing on the method and being tested for different time series to explore more limitations and strengths.

For an epidemiological analysis, studying the cycles of dengue serotypes is more interesting than the dataset since the circulating serotype varies over the years [12]. This contributes to a serotype that can stay longer without manifesting in a population with a large number of immune. At the same time, a different serotype may return soon after an outbreak of another serotype, contributing to subsequent epidemics, and it can significantly impact disease dynamics.

4 Conclusions

The Visibility Graph and Horizontal Visibility Graph provided distinct insights into dengue data for Rio de Janeiro and Belo Horizonte, demonstrating that the disease dynamics vary between cities. This variability highlights the need for studies involving more cities to explore further and refine the technique. Additionally, analyzing dengue cases by serotype could offer more valuable insights into the recurrence of the disease within populations.

The impact-frequency histogram introduced in this study offers a novel tool for assessing disease periodicity. We hope that this technique will support epidemiological analyses and enhance our understanding of the dynamics of dengue and other cyclical diseases.

References

1. Wong, C.: Climate change is also a health crisis-these 3 graphics explain why. Nature (2023)
2. de Carvalho Araújo, F.M., Nogueira, R.M.R., de Araújo, J.M.G., Ramalho, I.L.C., de Sá Roriz, M.L.F., de Melo, M.E.L., Coelho, I.C.B.: Concurrent infection with dengue virus type-2 and DENV-3 in a patient from Ceará, Brazil. Memórias do Instituto Oswaldo Cruz **101**(8), 925–928 (2006)
3. de Mattos Almeida, M.C., Assunção, R.M., Proietti, F.A., Caiaffa, W.T.: Intra-urban dynamics of dengue epidemics in Belo Horizonte, Minas Gerais State, Brazil, 1996–2002. Cadernos de Saúde Pública **24**(10) 2385–2395 (2008)
4. Barreto, M.L., Teixeira, M.G.: Dengue no Brasil: situação epidemiológica e contribuições para uma agenda de pesquisa. Estudos Avançados **22**(64), 53–72 (2008)
5. da Consolação Magalhães Cunha, M., Caiaffa, W.T., di Lorenzo Oliveira, C., Kroon, E.G., Pessanha, J.E.M., Lima, J.A., Proietti, F.A.: Fatores associados à infecção pelo vírus do dengue no Município de Belo Horizonte, Estado de Minas Gerais, Brasil: características individuais e diferenças intra-urbanas. Epidemiologia e serviços de saúde **17**(3), 217–230 (2008)
6. Gubler, D.J.: Cities spawn epidemic dengue viruses. Nat. Med. **10**(2), 129–130 (2004)
7. Coelho, F.C., De Carvalho, L.M.: Estimating the attack ratio of dengue epidemics under time-varying force of infection using aggregated notification data. Sci. Rep. **5** (2015)
8. Donalísio, M.R., Glasser, C.M.: Vigilância entomológica e controle de vetores do dengue. Rev. bras. epidemiol **5**(3), 259–272 (2002)
9. Lima, T.F.M., Lana, R.M., de Senna Carneiro, T.G., Codeço, C.T., Machado, G.S., Ferreira, L.S., de Castro Medeiros, L.C., Davis Junior, C.A., DengueME: a tool for the modeling and simulation of dengue spatiotemporal dynamics, Int. J. Environ. Res. Pub. Health **13**(9), 920 (2016)
10. Reich, N.G., Shrestha, S., King, A.A., Rohani, P., Lessler, J., Kalayanarooj, S., Yoon, I.-K., Gibbons, R.V., Burke, D.S., Cummings, D.A.: Interactions between serotypes of dengue highlight epidemiological impact of cross-immunity. J. Royal Soc. Interface **10**(86), 20130414 (2013)
11. Borchering, R.K., Huang, A.T., Mier-y Teran-Romero, L., Rojas, D.P., Rodriguez-Barraquer, I., Katzelnick, L.C., Martinez, S.D., King, G.D., Cinkovich, S.C., Lessler, J., et al.: Impacts of zika emergence in latin america on endemic dengue transmission. Nat. Commun. **10**(1), 1–9 (2019)
12. Ximenes, R., Amaku, M., Lopez, L.F., Coutinho, F., Burattini, M.N., Greenhalgh, D., Wilder-Smith, A., Struchiner, C.J., Massad, E.: The risk of dengue for non-immune foreign visitors to the 2016 Summer Olympic Games in Rio de Janeiro, Brazil. BMC Infect. Dis. **16**(1), 186 (2016)
13. Siriyasatien, P., Chadsuthi, S., Jampachaisri, K., Kesorn, K.: Dengue epidemics prediction: a survey of the state-of-the-art based on data science processes. IEEE Access **6**, 53757–53795 (2018)
14. Lacasa, L., Luque, B., Ballesteros, F., Luque, J., Nuno, J.C.: From time series to complex networks: the visibility graph. Proc. Natl. Acad. Sci. **105**(13), 4972–4975 (2008)
15. da Saúde (Brasil), M.: Sistema de informação de agravos de notificação (2019). http://portalsinan.saude.gov.br/
16. IBGE: Instituto brasileiro de geografia e estatístíca (2023). https://www.ibge.gov.br/
17. Campos, N.B.D., Morais, M.H.F., Ceolin, A.P.R., da Consolaçã o Magalhães Cunha, M., Nicolino, R.R., Schultes, O.L., de Lima Friche, A.A., Caiaffa, W.T.: Twenty-two years of dengue fever (1996–2017): an epidemiological study in a Brazilian city. Int. J. Environ. Health Res. **31**(3), 315–324 (2021)
18. de Oliveira Lemos, L., Júnior, A.C.O., de Assis Mendonça, F.: Urban climate maps as a public health tool for urban planning: the case of dengue fever in Rio de Janeiro, Brazil. Urban Clim. **35**, 100749 (2021)
19. Polwiang, S.: The time series seasonal patterns of dengue fever and associated weather variables in Bangkok (2003–2017). BMC Infect. Dis. **20**(1), 1–10 (2020)

20. Team, R.C., et al.: R: A language and environment for statistical computing (2013)
21. Damtew, Y.T., Tong, M., Varghese, B.M., Anikeeva, O., Hansen, A., Dear, K., Zhang, Y., Morgan, G., Driscoll, T., Capon, T., et al.: Effects of high temperatures and heatwaves on dengue fever: a systematic review and meta-analysis. EBioMedicine **91** (2023)
22. Chien, L.C., Yu, H.L.: Impact of meteorological factors on the spatiotemporal patterns of dengue fever incidence. Environ. Int. **73**, 46–56 (2014)

The Brazilian Maritime Network During the COVID-19 Pandemic: Analysis of Topologies and Impacts on Connectivity

Carlos César Ribeiro Santos, Hernane Borges de Barros Pereira, Thiago Barros Murari, Leonardo Sanches de Carvalho Filho, and Marcelo do Vale Cunha

Abstract This article examines the structural and topological transformations in the Brazilian maritime network during the COVID-19 pandemic, with a focus on adaptations made to maintain the efficiency, connectivity and resilience of cargo transport in the country. In a pandemic context where global trade was impacted by restrictions and shifts in supply and demand, complex network theory was used to analyze the structure and properties of this network. The methodology is based on quantitative analysis of data from the Automatic Identification System (AIS), which monitors the routes and movements of ships between Brazilian ports in the years 2019 (pre pandemic) and 2020 (during the pandemic). Topological metrics such as centrality, modularity, the clustering coefficient, and the average path length within the network are evaluated. The results indicate significant changes, with the strengthening of regional hubs, such as the ports of Manaus and Suape, and a redistribution of cargo flows that created denser regional clusters. The analysis also reveals a decreased in network modularity and a reduction in small world characteristics, resulting in greater average distances between ports and higher time and logistical costs for long-distance routes.

Keywords Maritime network · COVID-19 · Brazil

C. C. R. Santos (✉) · H. B. de Barros Pereira · T. B. Murari · L. S. de Carvalho Filho
SENAI CIMATEC University, Salvador, BA, Brazil
e-mail: mailto.carlos.santos@fieb.org.br; carlos.santos@fieb.org.br

T. B. Murari
e-mail: thiago.murari@fieb.org.br

M. do Vale Cunha
Instituto Federal da Bahia, Barreiras, BA, Brazil

M. Macedo et al. (eds.), *Complex Networks XVI*, Springer Proceedings in Complexity,
https://doi.org/10.1007/978-3-031-93619-7_8

1 Introduction

Organizations operate in a highly competitive, dynamic, complex, and unstable global environment. These characteristics denote unpredictable scenarios in which all operations and activities must be continuously reviewed. In this business context, it is crucial for organizations to seek effective strategies that can enhance operational efficiency, particularly in terms of physical distribution. From this perspective, logistics is increasingly considered a key component in generating competitive advantages for nations and companies, ensuring the delivery of goods and services at the right time and place, under the desired conditions, and at the lowest possible cost. Consequently, the technical study of transportation modes has become increasingly relevant for the global logistics chain.

The COVID-19 pandemic has underscored the crucial role of resilience in maritime networks, particularly in the container transport sector. As a dominant means of global supply chain flow, container transport faced significant disruptions due to health-related restrictions at ports, border crossings, and inland transport sectors. This resulted in blank sailings, delays, and inflated freight rates, which exposed vulnerabilities in the maritime network, emphasizing the importance of flexible and adaptive strategies to mitigate disruptions and maintain connectivity across critical trade routes [1, 2].

Resilience in maritime logistics can be assessed through complex network theory, which helps analyze network structure, connectivity, and vulnerability. Recent studies have shown that large and well-connected ports generally withstand crises more effectively, while smaller ports serving as transshipment hubs or regional bridges are more susceptible to disruptions. The geographical positioning of ports also influences resilience, as ports with strategic locations have a greater capacity to absorb shocks and redirect flows when needed. This geographic factor is increasingly integrated into network vulnerability models to create more accurate and realistic assessments [3, 4].

The pandemic highlighted the importance of diversified regional hubs within maritime networks. For example, some regions adapted to logistical bottlenecks by shifting flows to secondary ports, allowing regional clusters to take on greater significance. This shift not only maintained cargo movement but also demonstrated how localized clusters can enhance resilience against widespread network disruptions. However, this adaptation came with higher logistical costs due to the longer distances and more indirect routes involved, reflecting a trade-off between resilience and efficiency in crisis contexts [5].

Specifically, in Brazil, the movement of cargo via maritime transportation has grown exponentially over the past twenty years due to the country's commercial and financial liberalization. Globally, approximately 90% of cargo transportation volume in trade is conducted through maritime routes, making this mode essential for the competitive development of countries and regions. For Brazil, with an extensive 7,491 km coastline—the 16th longest in the world—and an economy heavily reliant on commodity exports, the maritime network is vital to the national economy, linking ports in various regions and facilitating integration with global markets. A 2022 study by the National Confederation of Transport (CNT) on the basis of data from the National Waterway Transportation Agency (Antaq) revealed that Brazil utilizes only approximately 30% of its navigable waterways, highlighting the untapped logistical potential of this transportation mode.

However, the COVID-19 pandemic has exposed significant vulnerabilities in maritime logistics systems worldwide, including the Brazilian port system. Owing to increased travel restrictions, reduced port activity, and changes in supply and demand, maritime networks face significant impacts, resulting in delays, bottlenecks, and a pressing need for quick and effective adaptations.

In this uncertain scenario, analyzing the structure and dynamics of the Brazilian maritime network during the pandemic is essential for understanding how the system responded to this global crisis. Complex network theory provides a valuable theoretical framework for exploring the structural properties of this network, enabling an assessment of port resilience, connectivity, and adaptability. Specifically, identifying and analyzing different network topologies, such as scale-free, small-world, and hierarchical networks, can offer deep insights into how the network reorganized to maintain cargo flow despite pandemic-related restrictions.

This paper aims to analyze the structural and topological changes in the Brazilian maritime network during the COVID-19 pandemic, focusing on connectivity, efficiency, and resilience properties, as well as the adaptations made to ensure cargo movement. The specific objectives are as follows:

- To map the Brazilian maritime network before and during the COVID-19 pandemic, major connectivity changes between ports and route adjustments were identified.
- To evaluate the network's topological properties, including centrality, modularity, the clustering coefficient, and average path length, changes in and the implications of each metric over time are highlighted.
- To analyze regional cluster formation and flow redistribution between central hubs and regional ports, how this adaptation contributes to the network's operational resilience must be identified.

Studying maritime networks from a complex network perspective is particularly relevant today, as goods transportation faces significant challenges in terms of security, efficiency, and continuity. The COVID-19 pandemic highlighted critical vulnerabilities in the Brazilian port system, underscoring the importance of understanding the resilience of logistics and transport networks during large-scale crises. By analyzing the topologies and structure of the Brazilian maritime network, it is possible to identify areas for improvement to enhance the system's robustness and adaptability in the face of future crises.

Moreover, increased regionalization of cargo flows and the redistribution of centrality to secondary hubs may provide valuable lessons for formulating policies and investment strategies in port infrastructure. These measures aim to optimize maritime transportation efficiency and enhance the network's responsiveness and adaptability to potential future disruptions. Thus, the analysis presented in this paper contributes to developing a safer, more efficient, and resilient maritime network for the Brazilian waterway transport sector.

2 Background

Studying maritime networks from a complex network perspective has gained increasing attention in the literature because of the strategic importance of this sector for international trade and the need for a deep understanding of its resilience and efficiency. Complex network theory provides robust analytical tools for modeling and analyzing maritime networks' structural and dynamic properties, enabling the assessment of metrics such as centrality, modularity, and the clustering coefficient, which are essential for understanding network connectivity and vulnerability.

2.1 *Maritime Networks*

According to Wasserman and Faust [6], a social network is formed by one or more types of relationships between real-world objects (e.g., people, institutions). Network vertices are called actors, and the social relationships connecting these actors are called edges. The structure of a social network can be modeled by a graph $G = (V, E)$, where V is the set of vertices containing n elements, and E, with m elements, is the set of edges, i.e., pairs of vertices that are related through some preestablished criteria. The network's topological characterization and community comparisons can be achieved through statistical indices based solely on the information contained in these two sets.

The maritime network presented here, as defined by Santos et al. [7], represents the relationship between ports (vertices) and their respective ship movements (edges) along the Brazilian coast during a specified period, which serves as a parameter for this research.

2.2 *Complex Networks and Maritime Transport Topologies*

Authors such as Ducruet and Notteboom [8] have explored global maritime network topologies, arguing that maritime transport networks exhibit a scale-free structure in which a few ports serve as highly connected hubs, whereas most ports have limited connectivity. This structure makes maritime networks robust to random failures but vulnerable to disruptions at main hubs, as demonstrated during the COVID-19 pandemic. Their work emphasized the importance of understanding these topological features to develop mitigation strategies that ensure maritime transport continuity in times of crisis.

Other studies like [9] examining maritime network resilience and the impact of disruptive events, such as the pandemic, on port connectivity. They highlight the importance of route diversification and strengthening regional hubs to reduce dependency on major hubs and increase network resilience. This approach offers valuable insights for Brazil, where regional hub diversification was necessary to mitigate the impact of pandemic restrictions.

2.3 Regionalization and Adaptation in Maritime Networks

Kanrak et al. [10] observed regionalization in the maritime network globally, reflecting a reorganization of routes into local clusters that allowed trade continuity despite global constraints. This regional cluster analysis is crucial for the Brazilian network, where regional subnetworks, with ports such as Manaus and Suape assuming local hub roles, contributed to maintaining cargo movement during the health crisis.

2.4 Network Metrics and Topological Properties

To evaluate the structure and behavior of maritime networks, various metrics are used, such as degree centrality and betweenness centrality, which are vital for maritime networks and indicate a port's ability to act as a connection point among different routes. Lam and Yap [11] stress the relevance of this metric in identifying critical hubs and network vulnerabilities.

Ducruet and Notteboom [8] noted that a network with a low average distance between ports enhances cargo movement efficiency. COVID-19 increased this metric, reflecting decentralization and reliance on longer, indirect routes.

These theoretical foundations emphasize the importance of understanding maritime network structures to develop more resilient and robust transport systems. The contributions of authors such as Ducruet and Notteboom [8] provide a solid basis for analyzing the Brazilian network, especially in crisis contexts such as the COVID-19 pandemic.

3 Materials and Methods

This study's methodology is based on a quantitative and topological analysis of the Brazilian maritime network before and during the COVID-19 pandemic, using ship movement data between Brazilian ports for 2019 and 2020. Data from the automatic identification system (AIS), a globally used technology for real-time vessel monitoring, were collected for this purpose.

Table 1 Network metrics (2019 vs 2020)

Metrics	2019	2020
Average degree centrality	0.45	0.39
Average betweenness centrality	0.32	0.27
Clustering coefficient	0.47	0.53
Average shortest path length	2.9	3.4

The AIS tracks ships via transponders, with each vessel transmitting critical information such as location, speed, direction, identification, and destination. Satellite and coastal stations capture these data, which are used for traffic monitoring, safety, and data collection for studies such as this. AIS data are reliable and widely used for maritime network analysis, as they provide detailed records of ship routes and movements over time. The analysis steps for constructing the networks included the following steps:

1. **Data collection and structuring**: AIS data were collected for the years 2019 and 2020, capturing ship routes and traffic flows between Brazilian ports. These data include both domestic and international traffic, allowing a comprehensive analysis of the maritime network in terms of connectivity and cargo movement. After collection, the data were organized into a network model, where each port represented a node, and connections between ports (frequent shipping routes) were edges. Each edge's weight was defined on the basis of traffic volume, allowing for an accurate analysis of each connection's significance.
2. **Maritime Network Construction**: Networks were built for the 2019 (prepandemic) and 2020 (pandemic) periods, reflecting connectivity changes between Brazilian ports. Using complex network theory, topological metrics such as centrality, modularity, the clustering coefficient, and the average distance between ports were highlighted.
3. **Topological Metric Analysis**: Key metrics were calculated (see Table 1) to assess the structural properties of the Brazilian maritime network, including the following:

   - Degree Centrality and Betweenness Centrality: Identifying major hubs and assessing the central roles of certain ports before and during the pandemic;
   - Clustering Coefficient: To measure the level of interconnection among ports and identify regional clusters;
   - Average path length: To evaluate the connectivity efficiency between ports, especially considering the impact of regionalized routes in 2020;
   - Network comparison and visualization: The 2019 and 2020 networks were visualized and compared to identify structural and regional changes, modularity and local cluster formation during the pandemic. These visualizations emphasize key hubs and connectivity redistribution among ports.

This methodology enabled a detailed analysis of the changes in the Brazilian maritime network due to COVID-19, providing empirical insights into structural and logistical adaptations during the health crisis. This analysis is further discussed in the Data Analysis section.

4 Results and Discussion

First, it is important to say that considering that the networks and metrics analyzed in detail here are from the years 2019 and 2020, this research also examined years prior to 2019 and found that the trends observed from 2016 to 2019 reveal a period of stability and gradual improvement in the Brazilian maritime network. Key ports such as Santos and Paranaguá consistently held dominant positions, with steady increases in degree centrality and betweenness centrality. The clustering coefficient and average shortest path length indicate a network that was becoming increasingly efficient and interconnected.

However, the sharp deviations in 2020 across all metrics highlight the transformative effects of the COVID-19 pandemic. The decline in degree and betweenness centrality, coupled with the rise in the clustering coefficient and average shortest path length, underscores a shift toward regionalization and fragmentation. These changes were driven by the need to adapt to restrictions and logistical challenges, such as port closures and disruptions in global trade routes.

An analysis of Figs. 1 and 2 (representing the years 2019 and 2020, respectively) allows us to identify the main ports with the highest connectivity in the network, which act as central points for cargo movement and distribution.

The weight of each edge in the network was defined based on the "volume" of maritime traffic. In this study, volume refers to the number of operations (voyages carried out between two ports) during the 2019 and 2020 periods. This metric was

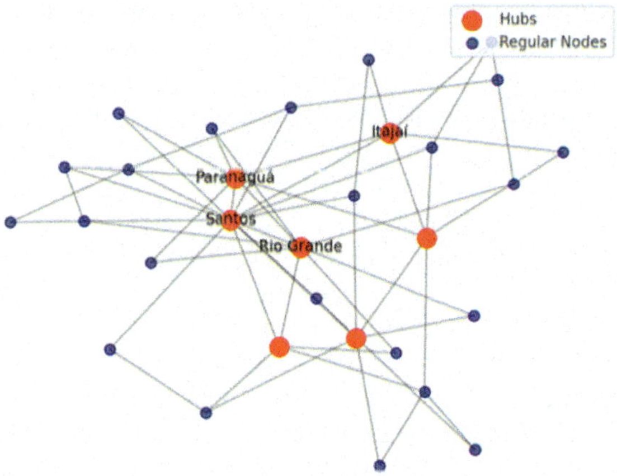

Fig. 1 Structure of the Brazilian maritime network (Year: 2019)

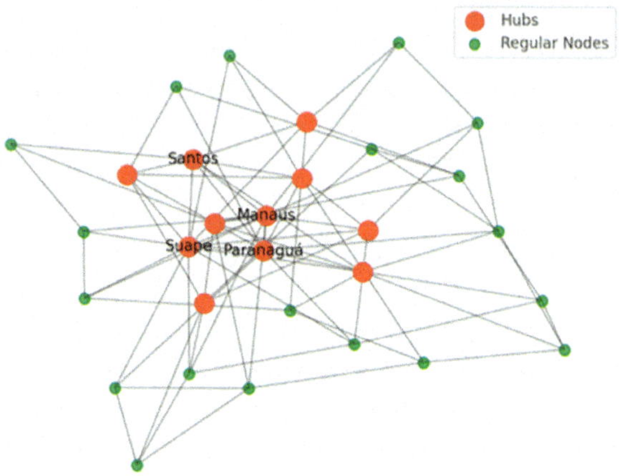

Fig. 2 Structure of the Brazilian maritime network (Year: 2020)

chosen for its ability to directly reflect the frequency of connections between ports, being less sensitive to fluctuations in transported weight or monetary value. This ensures a more robust and consistent analysis of changes in network connectivity and behavior.

The analysis confirmed that the network is fully connected, meaning all nodes (ports) have at least one direct connection to other ports. No isolated nodes were identified in the analyzed structure, corroborating the visual structure presented. The analysis confirms that the reported values of average degree centrality and average degree are consistent with their definitions for unweighted and connected networks. The network's structure, with no isolated components, reflects a functional and integrated system. Moreover, the equivalence between these metrics ensures that the classification of hubs and regional nodes aligns with the network's topological properties.

On the basis of the analysis of Fig. 1, we can identify the main hubs of Brazilian maritime navigation prior to COVID-19 in 2019, namely, the Port of Santos (SP), which is considered the largest port in Latin America, centralizing a large portion of Brazil's exports and imports; the Port of Paranaguá (PR), one of the main ports for the outflow of grains and agricultural products; the Port of Rio Grande (RS), which is significant for the trade of agricultural products, especially soybeans; and the Port of Itajaí (SC), one of the largest ports for container movement.

Figure 2 (2020) shows that the Port of Santos (SP) retained its central role during the COVID-19 pandemic, although it experienced delays due to health restrictions. The port of Paranaguá (PR) remained one of the primary outflow points, adapting its operations to limit gatherings. The role of the port of Manaus (AM) increased as a regional hub, as the isolation of certain regions encouraged the use of local hubs, and the port of Suape (PE) emerged as a significant hub in Northeast China, with its relevance increasing to meet regional demand.

The Brazilian maritime networks of 2019 and 2020 showed significant structural changes due to the impact of the COVID-19 pandemic. The main observed changes, beyond shifts in hub ports, include alterations in the connectivity distribution, the formation of regional clusters, and the topological properties of the network. Below, we highlight the main changes and analyze the network's topological properties.

4.1 Main Changes Identified

In 2019, the Brazilian maritime network was characterized by a more connected structure, with major hubs such as Santos and Paranaguá centralizing most routes. In 2020, logistical restrictions imposed by the pandemic led to a redistribution that increased the importance of regional hubs, such as Manaus and Suape, forming denser regional clusters. This resulted in fewer direct routes between major ports across different Brazilian regions, thereby reducing the network's efficiency in terms of global connectivity.

The pandemic also led to a decreased in network modularity, indicating a breakdown in the network's community structure. This reduction suggests that the COVID-19 pandemic led to a more homogeneous network, with logistical flows redistributed and a diminished presence of well-defined clusters.

The 2019 network displayed small-world topological characteristics, with short average distances between nodes, facilitating rapid good movement. In 2020, this characteristic diminished due to an increase in the average distance between ports caused by fewer direct routes between distant regions. This impact was especially noticeable between ports in Southeast China and other regions, hindering rapid connectivity across geographically distinct areas. Although hubs such as Santos and Paranaguá continued to play central roles, the network in 2020 showed greater adaptability by partially redistributing operations to regional ports. This resilience was essential to mitigate the impact of closures or congestion at major hubs.

It is important to say that in 2019, 12,000 operations were carried out, transporting approximately 250 million tons, with an average of 33 operations per day. In 2020, 10,500 operations were carried out, representing a 12.5% reduction due to the pandemic, but with an increase in the diversity of regional destinations. The average volume transported per operation decreased to 20 million tons per quarter, and the daily average fell to 29 operations. The Port of Manaus showed a 15% increase in the number of operations in 2020, while the Port of Santos experienced an 8% reduction. The average distance traveled between ports increased by 18%, indicating greater fragmentation and regionalization of routes.

4.2 Analysis of Topological Properties

To understand these changes in greater depth, let us examine the key topological properties. In 2019, ports such as Santos and Paranaguá had high degree centrality, serving as main nodes for cargo distribution. With the pandemic in 2020, the degree centrality of these ports decreased slightly, whereas ports such as Manaus and Suape gained greater centrality because of increased regional connections. This redistribution indicates a reduced traffic concentration at a single point, promoting a more diversified and resilient network.

The betweenness centrality of major hubs (especially Santos and Paranaguá) declined in 2020. This shift occurred because the network became less dependent on direct connections between these hubs, focusing instead on regional routes. The reduction in betweenness centrality at key hubs suggests that the network adapted to prevent congestion at these points by increasing the importance of regional ports.

The clustering coefficient, which measures the degree of connectivity among a node's neighbors, increased in the 2020 network because of the formation of denser regional clusters. This indicates that the ports within each cluster became more interconnected. This heightened regional clustering benefited internal logistics, facilitating the circulation of goods at the regional level, although it compromised the efficiency of exchanges at the national level.

The average shortest path length between nodes increased in 2020, reflecting a reduction in the number of direct routes between different regions. This change implies higher transportation times and logistics costs for long-distance routes. In practical terms, this means that the 2020 network became more reliant on indirect routes, with multiple intermediate ports needed to reach more distant destinations.

4.3 Implications of Changes in Topological Properties

The changes in the Brazilian maritime network's topological properties during the pandemic had several implications, including higher transportation costs and times due to increased average distances between ports and fewer direct routes, impacting logistical efficiency. Additionally, the formation of regional clusters made the network more resilient, allowing operations to continue at the local level even with restrictions at larger ports. This adaptation is valuable for future crises, as the network can maintain functionality on a more localized scale.

The lower betweenness centrality at major hubs suggests that the network adapted to mitigate the risk of congestion at critical ports, redistributing demand to other regional hubs. This behavior reinforces the importance of developing and maintaining a diversified port infrastructure.

These observations are essential for future logistical planning in Brazil, especially regarding the construction of a more decentralized and robust port network to withstand potential future crises.

5 Conclusions

The identification and role of hubs within Brazilian maritime networks have several technical implications, especially during crises such as the COVID-19 pandemic. Ports such as Santos and Paranaguá handle a significant volume of cargo movement, making them vulnerable to overload during times of crisis. With restrictions and health measures imposed during the pandemic, these hubs experienced delays, impacting the logistics chain and increasing transportation times.

Additionally, during the COVID-19 pandemic, increased regionalization elevated the role of local hubs, such as Manaus and Suape, which took on greater prominence within their respective regions. This shift helped alleviate the logistical challenges of long-distance transport but also led to a redistribution that, in some cases, increased transportation costs and times. The maritime network demonstrated a degree of resilience, as smaller ports were able to partially absorb the flow that would typically be handled by major hubs. This redistribution highlights the importance of multiple regional hubs to ensure network functionality in the face of disruptions, such as pandemic-related restrictions.

In conclusion, the analysis of the Brazilian maritime network during the COVID-19 pandemic revealed that despite reduced connectivity and the formation of regional clusters, the network was able to adapt to the new demands and restrictions imposed by the pandemic. The network's scale-free structure contributed to its resilience, allowing other ports to assume part of the operations when major hubs faced restrictions. However, the loss of small-world characteristics increased the average distance between ports, reducing the efficiency of goods transport.

These findings provide valuable insights for policy-making aimed at enhancing the resilience of the Brazilian maritime network, reinforcing the importance of secondary hubs, and promoting regionalization strategies to mitigate the impacts of future crises.

References

1. Guerrero, D., Letrouit, L., Pais-Montes, C.: The container transport system during Covid-19: an analysis through the prism of complex networks. Transp. Policy **115**, 113–125 (2022)
2. Notteboom, T., Pallis, T., Rodrigue, J.P.: Disruptions and resilience in global container shipping and ports: the COVID-19 pandemic versus the 2008–2009 financial crisis. Marit. Econ. Logist. **23**(2), 179 (2021)
3. Álvarez, N.G., Adenso-Díaz, B., Calzada-Infante, L.: Maritime traffic as a complex network: a systematic review. Netw. Spat. Econ. **21**(2), 387–417 (2021)
4. Guo, J., Feng, T., Wang, S., Qin, Y., Yu, X.: Shipping network vulnerability assessment integrated with geographical locations. Transp. Res. Part D Transp. Environ. **130**, 104166 (2024)
5. Kuźmicz, K.A.: Impact of the COVID-19 pandemic disruptions on container transport. Eng. Manag. Prod. Serv. **14**(2), 106–115 (2022)
6. Wasserman, S., Faust, K.: Social Network Analysis. [S.l.]. Cambridge University Press, Cambridge (1994)

7. Santos, C.C.R., De Barros Pereira, H.B., Da Silva Palmeira, A., Do Vale Cunha, M.: Aplicação Da Teoria De Redes Para Análise Logística Dos Hubports Da Cabotagem Brasileira. Revista Mundi Engenharia, Tecnologia e Gestão (ISSN: 2525–4782). **4**, 165-1–165-18 (2019)
8. Ducruet, C., Notteboom, T.: The Worldwide Maritime Network and Its Regional Dynamics. Global Networks, Wiley (2012)
9. Dirzka, C., Acciaro, M.: Global shipping network dynamics during the COVID-19 pandemic's initial phases. J. Transp. Geogr. **99** (2022)
10. Kanrak, M., Nguyen, H.-O., Du, Y.: Maritime transport network analysis: a critical review of analytical methods and applications. J. Int. Logist. Trade (2019)
11. Lam, J.S.L., Yap, W.Y.: Dynamics of liner shipping network and port connectivity in supply chain systems: analysis on East Asia. J. Transp. Geogr. **19**(6), 1272–1281 (2011)

Understanding the Structure and Resilience of the Brazilian Federal Road Network Through Network Science

Júlio Taveira, Fernando Buarque, and Ronaldo Menezes

Abstract Understanding how transportation networks work is important for improving connectivity, efficiency, and safety. In Brazil, where road transport is a significant portion of freight and passenger movement, network science can provide valuable insights into the structural properties of the infrastructure, thus helping decision makers responsible for proposing improvements to the system. This paper models the federal road network as weighted networks, with the intent to unveil its topological characteristics and identify key locations (cities) that play important roles for the country through 75,000 km of roads. We start with a simple network to examine basic connectivity and topology, where weights are the distance of the road segment. We then incorporate other weights representing number of incidents, population, and number of cities in-between each segment. We then focus on community detection as a way to identify clusters of cities that form cohesive groups within a network. Our findings aim to bring clarity to the overall structure of federal roads in Brazil, thus providing actionable insights for improving infrastructure planning and prioritising resources to enhance network resilience.

Keywords Complex networks · Road network · Network science · Resilience · Brazil

J. Taveira · F. Buarque
University of Pernambuco, Recife, Brazil

J. Taveira (✉)
Federal Highway Police University, Florianópolis, Brazil
e-mail: jcft@ecomp.poli.br

R. Menezes
University of Exeter, Exeter, UK

M. Macedo et al. (eds.), *Complex Networks XVI*, Springer Proceedings in Complexity,
https://doi.org/10.1007/978-3-031-93619-7_9

9.1 Introduction

In the global economy, road infrastructures are a fundamental component because they are key in facilitating the movement of goods and people. Many countries rely heavily on road networks for the transport of commodities, with some nations being more dependent than others. For instance, in geographically vast countries with dispersed urban centres, road transport becomes essential for economic activities and regional connectivity.[1]

Brazil's dependence on roads is particularly significant, serving as the primary mode of transportation for freight (and passengers) across its extensive, and often difficult access, territory. The federal highways (henceforth called "roads", given that they are not always highways, i.e., multiple lanes) play a crucial role in the development and integration of smaller cities. It has been shown that the efficiency and reliability of roads directly impact economic growth, social cohesion, and access to services [17, 24].

Brazil's has a specific police force responsible for federal roads: the Federal Highway Police (Polícia Rodoviária Federal, PRF). They are responsible for about 75,000 km of roads out of the nearly 2 million km total [14]. While this sounds small compared to the total, they are the main arteries of road transportation in Brazil; 60% of goods are transported using the federal road structure making its connectivity [8] crucial to the country.

According to the Brazilian constitution [16], the PRF is responsible for patrolling federal roads, enforcing traffic laws, and ensuring the road safety. Their duties include accident prevention, combating criminal activities, and providing assistance to motorists, which are vital for maintaining the operational integrity of the nation's road network, among others. While the PRF is very knowledgeable about the road infrastructure, this knowledge is often distributed, not providing a holistic view of the infrastructure. By analysing network structures, authorities can identify critical nodes and links that require investment, improvement, or even more policing.

While the construction of new federal roads is relatively infrequent in Brazil (the country continues to depend heavily on infrastructure investments made in the1960s and 1970s [8], see Sect. 9.3), understanding the structure of the existing network is paramount. The network's structural properties are influenced by factors such as human mobility patterns [4], population in neighbouring cities, and various events including festivals, religious activities, holidays, and freight logistic decisions. These factors can lead to changes in the significance and utilisation of specific roads, highlighting the importance of a comprehensive structural analysis. In this study, we model the Brazilian federal road network using the 546 largest cities as nodes; medium and large cities as per the Brazilian Geography and Statistics Institute (IBGE) [11].

[1] Five of the six largest countries in area also rank among the top five in road network size. Source: https://en.wikipedia.org/wiki/List_of_countries_by_road_network_size (accessed: 25 October 2024).

9.2 Transportation Networks

Transportation networks are a critical aspect of modern society, and intrinsically linked to economic growth and social development. They facilitate the movement of goods, services, and people, thereby connecting markets and fostering globalisation [29]. The structure and dynamics of these networks is an essential part of the process of improving efficiency, resilience, and sustainability around the world.

Over the past few decades, the world has effectively become smaller due to advancements in transportation and communication technologies. This phenomenon, often referred to as the "time-space convergence," implies that the relative distance between places decreases as connectivity improves [20]. For example, the average transatlantic travel time for freight shipments in the 1800s was about 30 days, while today is about 8 h. While increased connectivity offers benefits, it also presents concerns. Faster transportation networks can facilitate the rapid spread of diseases, as evidenced by the global transmission of pandemics like COVID-19 [13]. Similarly, invasive species (e.g., plants, animals) can spread more easily through connected pathways, disrupting ecosystems and sometimes even local economies [21].

Network Science has been instrumental in modelling transportation networks, offering tools and methods to analyse their complex structures and dynamics [27]. Numerous studies have employed network theory to investigate various modes of transportation and spatial networks [6], providing insights into their topology, robustness, and vulnerability. For instance, air transportation networks have been extensively studied due to their global importance [19, 23]. Rail networks have also been a subject of interest with indications that these networks have small-world properties, identifying its structural characteristics and the implications for efficiency and robustness [25, 30]. In maritime transport, Kaluza et al. [22] studied the global cargo ship network, highlighting patterns in maritime traffic and their environmental impact.

In addition to air, rail and maritime studies, there has been significant work in urban road networks [1, 3, 12] but comparatively fewer investigations into country-wide road networks [33, 34], especially in the context of the Global South and large-scale countries. Road networks are particularly crucial in such regions due to their role in regional connectivity and economic development, where other forms of transport may be less developed or accessible.

In this paper, we focus on the Brazilian Federal Road Network (BFRN) using Network Science methodologies. By describing its structural properties, we aim to provide insights that can inform data-driven decisions for infrastructure development, policymaking, and strategic planning.

9.3 Data and Methods

Brazil is the largest country in South America and the fifth largest in the world, covering an area of approximately 8.5 million square kilometres. Due to its vast size and diverse geography, efficient transportation infrastructure is essential for economic development and national integration. However, Brazil lacks connectivity via rail and often many locations are not connected via airports, which make the road network crucial to many of the country's regions.

The federal roads in Brazil are under the jurisdiction of the federal government and serve as the main arteries linking major cities, ports, airports, strategic areas, and even to neighbouring countries. The primary purpose of these roads is to promote national connectivity, support economic activities, and ensure access to all areas of the country.

The development of the Brazilian road network happened mostly in mid-20th century, particularly during the 1950s and 1960s [31]. Brazil investment can be supported by economic theories which suggest that improved transportation infrastructure leads to reduced product costs, increased trade, and regional development [2].

The Brazilian federal road infrastructure comprises approximately 75,000 km of roads [9]. These are named according to a system that reflects their general direction and geographic location, as shown in Table 9.1. The roads are depicted in Fig. 9.1 (left), and the colours represent the different types of roads.

For modelling the proposed networks, we used several datasets. The road structure was obtained from the National Road System (NRS) [26], which provides georeferenced data for each kilometre of all Brazilian federal roads. From this dataset, we extracted information such latitude, longitude, and cities along the way. This dataset includes more than 130,000 geographic positions and was collected in 2022.

Table 9.1 Brazilian Federal Road naming conventions

Road type	General format	Colour	Example	Description
Radial	BR-0XX	—	BR-010, BR-020	Roads from Brasília to country's edges
Longitudinal	BR-1XX	—	BR-101, BR-116	Roads oriented N-S
Transversal	BR-2XX	—	BR-222, BR262	Roads oriented E-W roads
Diagonal	BR-3XX	—	BR-365, BR-319	Roads oriented NW-SE or NE-SW
Connecting	BR-4XX	—	BR-407, BR488	Generally connect federal roads

They are based on where they start in the country, their direction, and function [15]. The colours in the table refer to the map in Fig. 9.1 (left) where all the federal roads are depicted

Fig. 9.1 Brazilian Federal Road Network. We see on the (left) the picture showing the road types in different colours as defined in Table 9.1. However, some of these roads are not connected to others, so we use a modelling based on the 546 cities in Brazil and the federal roads that connected them leading to the giant component in the (centre). Last, we show the population distribution of Brazil (right); note the concentration around the east coast

We also incorporated information about all 5,570 cities in Brazil, including their populations as collected in 2022 by the Brazilian Institute of Geography and Statistics (IBGE) [11]. We utilised traffic incident data collected by the PRF, available through their open data portal [10]. This database contains over 485,000 incident records between 2017 and 2023, providing details such as date, time, number of people involved, injuries, location, to name a few.

For modelling the nodes, we selected cities classified by the IBGE as medium or large, with populations of at least 50,000 people [11]. Brazil has a total of 5,570 cities, of which 3,126 are intersected by a federal road. The 544 selected cities have approximately 125 million residents, representing 76% of the 164 million people residing in all cities traversed by federal road. In order to generate a consistent, we included an additional 30 small cities that are important for connecting roads (e.g., crossing points), resulting in a total of 574 cities. The nature of the Brazilian road network system, is that some of these cities are isolated and not part of the connected network of cities. Hence, we are left 546 cities which form the basis of the BFRN (and variations); the network giant component.

The edges in our network represent the roads linking the 546 selected cities. We utilised the NRS data, which contains information for each kilometre of road, to calculate the distances, number of cities, population, and incidents between pairs of cities. This calculation includes the end point as part of the weight; for instance, the population of people living between city A and city B, includes the population of A and B. Figure 9.1 (centre) illustrates the road network, more specifically the giant component of the network which we call BFRN.

We consider different weights in the structural analysis in this paper, giving rise to four weighted networks:

- BFRN: The road network using the distance as weight of the edges. This is essentially the network show in Fig. 9.1 (centre).

Table 9.2 Basic measures of the Brazilian Federal Road Network

Metric	BFRN	cBFRN	pBFRN	iBFRN
Number of nodes (n)	546	∘	∘	∘
Number of edges (m)	761	∘	∘	∘
Average degree ($\langle k \rangle$)	2.78	∘	∘	∘
Weighted distribution exponent (γ_w)	1.78	1.76	1.73	1.45
Average shortest path ($\langle \ell_w \rangle$)	1,784	75	4,324,144	130
Diameter (D_w)	5,768	213	22,902,144	45,659
Modularity (Q_w)	0.852	0.842	0.867	0.887

We see the values for the exponent of the power law weighted distribution for each network (γ_w), the average shortest path ($\langle \ell_w \rangle$), diameter (D_w), and the weighted modularity (Q_w). Decimal places are not displayed for $\langle \ell_w \rangle$ and D_w while for n, m, and $\langle k \rangle$ the values are shown once because they repeat for all networks

- cBFRN: The same nodes as BFRN but using the number of cities between the nodes as weight of the edges. For example, if between cities A and B we have N cities, the weight of the edge $w(A, B) = N + 2$ because the end points count for the weight.
- pBFRN: Weights here represent the total population in the cities in the cBFRN for that particular segment.
- iBFRN: Weights here represent the number of incidents that happened in-between the cities for the entire period of the study.

Table 9.2 shows some basic measurements on the four networks. The number of nodes (n), edges (m), and average degree ($\langle k \rangle$) do not change, and hence it is only shown for the BFRN. All the other measures are specific for each network.

9.4 Characterisation and Results

The goal in this paper is to characterise the Brazilian road network, leading to a better clarity of the infrastructure. Yet, there are contextual information that need to be mentioned because of the four weight variations used. In the BFRN the network can be used to understand hubs, effective paths, and more importantly how changes to this can lead to better efficiency in transportation times and fuel consumption; distance is highly correlated to travel time. In the cBFRN, one can concentrate on the identification of urban corridors as a function of the settlements, emergency planning, and planning of road maintenance making sure most cities remain connected at all times. A variation of this is the pBFRN where population of the cities are considered instead. In the pBFRN the analysis of paths (shortest) can demonstrate areas that are less populated and hence routes that may be maintained less often. Last, and relatively different from the others, is the iBFRN where the weights represent the number of incidents (i.e., deaths, injuries, driving under the influence). However, the specific applications described above fall outside the scope of this paper. Here we focus on identifying structural properties of these networks.

9.4.1 Weighted Degree Distributions

In spatial networks, nodes are embedded in physical space, and connections between them are constrained by geographical proximity. This spatial embedding imposes limitations on the number of connections a node can have, leading to degree distributions that are generally more homogeneous and do not follow a power-law distribution typically seen in other types of complex networks [6, 18].

When weights are assigned to the edges of spatial networks; representing attributes such as distances, the distribution of weighted degrees (also known as node strengths) can provide additional insights into the network's structure and functionality. With weighted edges, the degree distributions in spatial networks like road networks may follow power laws [5] defined as $P(k) = Ck^{-\gamma_w}$, where k represents the weighted degree of a node. For our networks, the power law fitting leads to values of $1 \leq \gamma_w \leq 2$ which shows that the networks are extremely hub-dominated, which is generally a sign of vulnerability.

In the context of the BFRN, examining the weighted degree distribution is essential for several reasons. When weights represent distances between cities, the distribution can highlight central nodes that are geographically significant due to their numerous or lengthy connections and perhaps some level of isolation given the other cities considered (with significant population) are far away. If weights represent the number of intermediate cities (cBFRN), the distribution uncovers nodes that are endpoints of urban corridors, indicating potential areas of high socio-economic activity or congestion. When considering weights as the total population served by each segment (pBFRN), the weighted degree distribution helps identify cities that are crucial for connecting large populations, or cities for which a large fraction of the population depend on the road as an economic drive. Lastly, with weights representing the number of incidents (iBFRN), the distribution can pinpoint nodes that are hotspots for accidents, guiding targeted safety interventions and resource allocation to ensure the police can quickly respond to these incidents.

Table 9.3 shows the 5 cities with the maximum and minimum weights. It is worth noting the 2 largest cities in Brazil are on the highly ranked in the pBFRN, which happens because of the artifact that their population counts for the weight of the edges they have. In fact, we see two other cities are part of the great São Paulo: Osasco and Guarulhos. Other points to observe is the correlation between the BFRN and cBFRN. We found that cities that have long distance roads attached to them are also

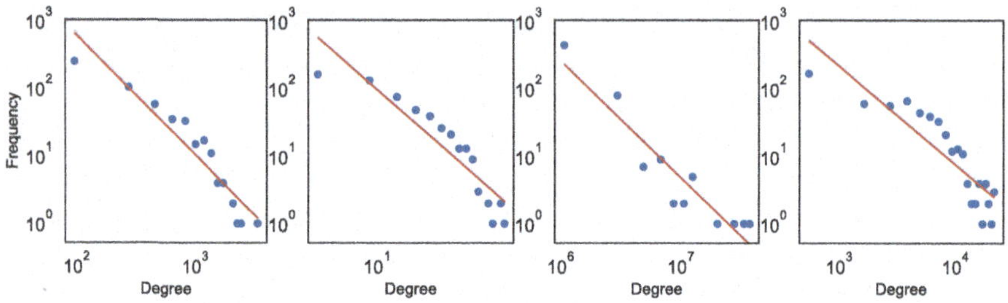

Fig. 9.2 Weighted degree distributions. The charts are for (from left to right): BFRN, cBFRN, pBFRN, and iBFRN. It's worth noting that the distributions follow a power law demonstrating super-hubs. However, except for the pBFRN [32], the networks appear to have some level of cut-off in their distribution which indicates limits to the values of the degrees

the ones with several intermediate cities. Last, we see that the cities featuring in high ranks in the iBFRN are not the ones with high population weight, leading to the idea that incidents may be related to other issues such as road quality and geographical accidents.

Overall, understanding the weighted degree distribution allows for a holistic analysis of the network's topology and its implications. It helps in identifying critical nodes whose failure or inefficiency could disproportionately affect the network's functionality. This knowledge is vital for enhancing network robustness, optimizing traffic flow, improving safety measures, and supporting strategic planning and policymaking aimed at improving the transportation infrastructure. Figure 9.2 shows the degree distribution for each of the four networks considered.

While individual cases for cities are important in decision-making, understanding the distribution of these degrees is also fundamental to a holistic view of the vulnerability. Table 9.3 shows that the networks appear to be super-hub dominated. The distributions are depicted in Fig. 9.2.

9.4.2 *Diameter and Paths*

The diameter of a weighted network, D_w is defined as the longest shortest path between any pair of nodes, taking into account the weights assigned to the edges, as below:

$$D_w = \max_{i,j \in n} d_{ij}^w,$$

where d_{ij}^w is the weighted shortest path length between cities i and j.

It represents the maximum cumulative weight that must be traversed to connect the most distant nodes in terms of the chosen weighting scheme. The diameter provides insights into the extent of the network's reach, potential bottlenecks, and areas where

Table 9.3 Top 5 cities with maximum and minimum weights

Rank	BFRN		cBFRN		pBFRN		iBFRN	
	Max	Min	Max	Min	Max	Min	Max	Min
1	Barreiras	Belém	Barreiras	Vitória	Osasco	Camocim	Registro	Rio Claro
2	Balsas	Vitória	Campo Mourão	Jandira	São Paulo	Três Passos	Piraquara	Brusque
3	Jacobina	Jandira	Jacobina	Aracaju	Rio de Janeiro	Caldas Novas	Betim	Valença
4	Gurupi	Itapevi	Balsas	Florianópolis	Guarulhos	Cametá	Brasília	Cabo Frio
5	Barra do Garças	Aracaju	Picos	Brusque	Brasília	Grajaú	Lages	Cametá

We can observe that, as expected, large metropolitan areas rank high in the pBFRN and they do not correlate directly to incidents in the iBFRN. However, there is some correlation to the list in BFRN and cBFRN saying that cities that have long distance weights are similar to the ones that have many cities as intermediate nodes connectivity may be limited or require improvement. The values for all the networks are shown in Table 9.2.

In the context of the BFRN, where the weights represent the distances between cities, the diameter signifies the greatest cumulative distance that must be covered to travel between the two most distant cities along the shortest possible route. This metric reflects the maximum physical separation in the network and can be used to assess the efficiency of national connectivity. Brazil, being a continental country with complex geographical features such as vast rivers, mountain ranges, and dense rainforests, inherently imposes lower bounds on these measures. For instance, the calculated diameter of the BFRN is 5,768 km, which exceeds the actual maximum straight-line distances across Brazil: 4,394 km from the northernmost point at the Monte Caburaí in Roraima to the southernmost point at Arroio Chuí in Rio Grande do Sul.

For the cBFRN, with weights representing the number of cities between each pair of connected cities, the diameter indicates the maximum number of cities that must be traversed along the shortest path between any two cities. This interpretation sheds light on the longest sequence of urban centres encountered on the most direct route, providing insights into urban density and regional development patterns. It can help identify corridors that connect numerous communities, potentially highlighting areas which could benefit from economic activity around transportation. For Brazil, there this diameter is 213 cities. This could be seen in conjunction to the actual population seen in the pBFRN, where weights denote the total population benefited by each segment, the diameter reflects the largest cumulative population connected along the shortest path between any two cities which for Brazil is about 22.9 million people which says that the diameter involves around 10% of the entire Brazilian population (214 million people).

Last, in the iBFRN, the diameter represents the highest cumulative number of incidents encountered along the safest (least incident-prone) path between any two cities. This metric identifies the pair of cities for which even the safest route involves traversing segments with a high total number of incidents, highlighting potential areas of concern for road safety. The cumulative value for this diameter is more than 45 thousand incidents; indicating that some of the roads are quite dangerous, with about 18 incidents per day for the 7-year period studied.

Figure 9.3 shows the diameters for the four networks studied using the colours defined in Table 9.1. Two things are worth noting, the significant difference between the BFRN and the iBFRN, with the latter being concentrated on the more populated area of the country. The second is that there is a strong correlation between the cBFRN and the pBFRN with the main difference being that the pBFRN naturally gravitates to the city of São Paulo given its large population.

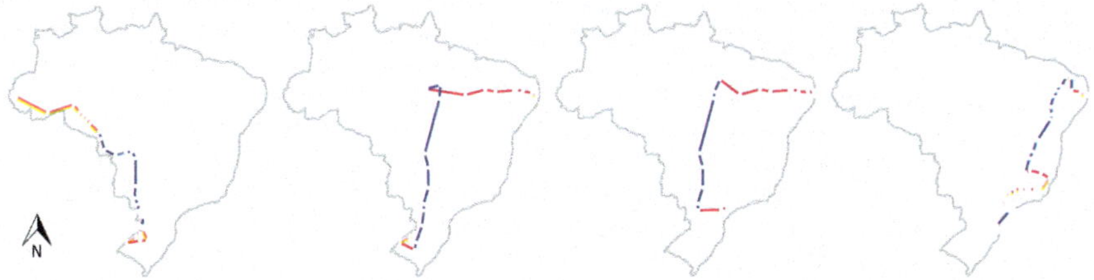

Fig. 9.3 Diameters of the networks. The diameters of the BFRN, cBFRN, pBFRN, and iBFRN (respectively left to right). The colours of the roads are used as defined in Table 9.1

9.4.3 Community Detection

Community detection is a fundamental aspect of network analysis, aiming to uncover the underlying structure of a network by identifying groups (communities) of nodes that are more densely connected internally than with the rest of the network. This process is essential for understanding the modular organisation of complex systems, revealing how entities interact within and across different subgroups.

Here we use the Louvain method [7] which is a widely used algorithm for community detection due to its efficiency and ability to handle large and weighted networks. It operates by maximising the modularity of the network, in our case the weighted modularity, Q_w, a measure that quantifies the quality of a particular division of the network into communities [28]. By considering edge weights, the Louvain method can detect communities that reflect not only the topology of the network but also the intensity of interactions between nodes (from weights).

In the case of the BFRN and variations, the communities may correspond to regions where there is a high volume of traffic, densely connected urban areas, or zones with significant safety concerns due to a high number of incidents. Figure 9.4 shows the 8 largest communities for the four networks, respectively from left to right: BFRN (19), iBFRN (26), cBFRN (20), pBFRN (20); the community analysis leads to more communities shown between parenthesis. One immediate aspect to observe is how the largest communities seem to be located in the coastal area of Brazil, where 55% of the population lives (within 150 km of the coast and 10% of total territory) [11].

It is worth noting that the community structure on the networks is quite strong, as shown by the modularity values in Table 9.3. Figure 9.4 depicts that the communities are quite regional, showing that the patterns of the nearby cities in the network have similarities.

Fig. 9.4 Network communities. For all networks, the communities show strong connections within regions. The picture shows the 8 largest communities for BFRN, cBFRN, pBFRN, and iBFRN (from left to right)

9.5 On Network Resilience

Resilience analysis, especially in the BFRN, is critical due to its vital role in connecting cities across Brazil's vast and diverse geography. The BFRN as the backbone of Brazil's transportation system, needs to show resilience to disconnections. Studying resilience allows for the identification of critical nodes (cities) whose removal, whether due to natural disasters, infrastructure failure, or other disruptions, could disproportionately impact the network. Natural and unnatural disasters, such as the devastating floods that recently impacted the southern region of Brazil, including the city of Porto Alegre, are likely to become more frequent and severe due to the growing effects of climate change. These events can lead to significant disruptions in critical infrastructure such as the BFRN highlighting the need for proactive resilience planning.

Here, we compare node removals based on degree, weighted degree, and betweenness to random removal for it brings clarity to the structural importance of different nodes in the BFRN. Degree, as a measure of the number of direct connections a node has, helps evaluate the impact of losing cities that are end points of several federal roads on the network's overall connectivity. Weighted degree extends this analysis by incorporating the significance of these connections, using weights such as distance, population, or the number of intermediate cities. Betweenness centrality, on the other hand, assesses the role of cities as intermediaries or bridges, measuring how often a node lies on the shortest paths between other cities. This highlights nodes that are critical for maintaining the global flow within the network. Random removal serves as a null model to compare how the network responds when disruptions occur without targeting specific structural properties.

The results shown in Fig. 9.5 (left) reveal that removing approximately 25% of the nodes based on degree leads to the complete destruction of the giant component, with faster destruction observed in the pBFRN. This outcome suggests that cities with a high number of connections are crucial for maintaining large-scale connectivity, especially when these connections serve densely populated areas. The earlier impact of weighted degree removal in the BFRN highlights that removing cities associated with large distances can disrupt connectivity also, but degree-based removals

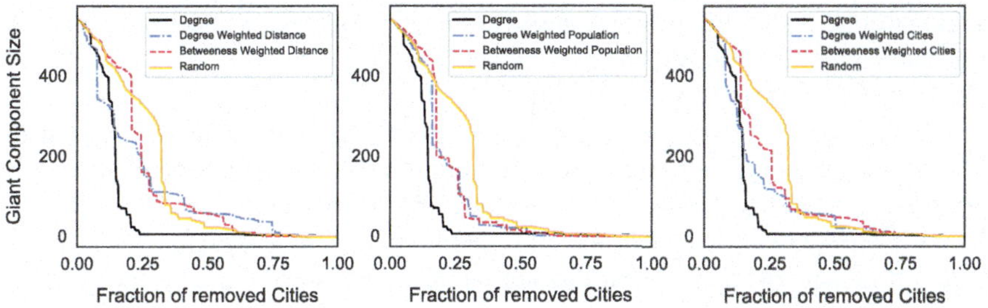

Fig. 9.5 Resilience analysis. The impact of node (city) removals based on degree, weighted degree, and betweenness, compared to random, highlights the vulnerability of the giant component particularly for the removal of highly connected cities (degree). Here we see BFRN, pBFRN and cBFRN respectively from left to right

ultimately have a stronger and more widespread effect, possibly because we are dealing with spatial networks. Interestingly, the faster disintegration caused by degree removals in the BFRN compared to betweenness suggests that, in spatial networks like the BFRN, direct connections may be more critical than the intermediary roles captured by betweenness. This is due to the physical constraints and cost associated with forming long-distance connections, which make high-degree nodes less likely but disproportionately important.

9.6 Discussion

This paper presented a comprehensive analysis of the BFRN by leveraging network science to evaluate its structure, connectivity, and vulnerabilities. Four distinct network models were considered, with weights based on distance, population, number of intermediate cities, and incidents. These models provided a multifaceted perspective on the BFRN, allowing for a holistic understanding of how different aspects of it contribute to its overall functionality. By analysing the structural properties of these weighted networks, we gained valuable insights into the key features that define the network's robustness and the critical nodes that play a pivotal role in maintaining connectivity.

The analysis highlights the importance of adopting a holistic approach to evaluate the BFRN, particularly in the context of decision-making by the Federal Highway Police (PRF). Limited resources require strategic priorities, and a comprehensive view of the network enables the identification of areas where interventions can have the greatest impact. For example, the pBFRN can guide decisions on resource allocation to highly populated regions, while the iBFRN informs strategies monitoring the segments and maybe making sure the incidents are not due to poor road maintenance. Similarly, cBFRN sheds light on the role of smaller urban areas in maintaining regional cohesion, offering opportunities to strengthen economic incentives to critical corridors that may otherwise be overlooked.

While resilience analysis demonstrated weaknesses associated with targeted node removals, the overarching value lies in understanding the broader structure of the BFRN. In the future, we could work to build on the foundation of this paper by integrating multilayer networks that include additional transportation modes, such as waterways, railways, and airports. Furthermore, the use of dynamic data, such as traffic flow patterns and natural disaster simulations, would enhance the ability to anticipate and mitigate potential disruptions.

References

1. Akbarzadeh, M., Memarmontazerin, S., Soleimani, S.: Where to look for power laws in urban road networks? Appl. Netw. Sci. **3**, 1–11 (2018)
2. Aschauer, D.A.: Is public expenditure productive? J. Monet. Econ. **23**(2), 177–200 (1989)
3. Badhrudeen, M., Derrible, S., Verma, T., Kermanshah, A., Furno, A.: A geometric classification of world urban road networks. Urban Sci. **6**(1), 11 (2022)
4. Barbosa, H., Barthelemy, M., Ghoshal, G., James, C.R., Lenormand, M., Louail, T., Menezes, R., Ramasco, J.J., Simini, F., Tomasini, M.: Human mobility: models and applications. Phys. Rep. **734**, 1–74 (2018)
5. Barrat, A., Barthelemy, M., Pastor-Satorras, R., Vespignani, A.: The architecture of complex weighted networks. Proc. Natl. Acad. Sci. **101**(11), 3747–3752 (2004)
6. Barthélemy, M.: Spatial networks. Phys. Rep. **499**(1–3), 1–101 (2011)
7. Blondel, V.D., Guillaume, J.L., Lambiotte, R., Lefebvre, E.: Fast unfolding of communities in large networks. J. Stat. Mech. Theory Exp. **2008**(10), P10,008 (2008)
8. Bottasso, A., Conti, M., de Sa Porto, P.C., Ferrari, C., Tei, A.: Roads to growth: the Brazilian way. Res. Transp. Econ. **90**, 101,086 (2021)
9. Brazilian Federal Highway Police (PRF): Annual report 2023. Technical report, Ministério da Justiça e Segurança Pública, Brasília, DF (2023). https://www.gov.br/prf/pt-br/acesso-a-informacao/dados-abertos/diest-arquivos/anuario-2023_final.html. Accessed 29 Oct 2024
10. Brazilian Federal Highway Police (PRF): Open data on traffic accidents from Brazilian federal highway police (2024). https://www.gov.br/prf/pt-br/acesso-a-informacao/dados-abertos/dados-abertos-acidentes. Ministry of Justice and Public Safety
11. Brazilian Institute of Geography and Statistics (IBGE): Brazilian population census (2022). https://censo2022.ibge.gov.br/
12. Chalkiadakis, C., Perdikouris, A., Vlahogianni, E.I.: Urban road network resilience metrics and their relationship: some experimental findings. Case Stud. Transp. Policy **10**(4), 2377–2392 (2022)
13. Chinazzi, M., Davis, J.T., Ajelli, M., Gioannini, C., Litvinova, M., Merler, S., Pastore y Piontti, A., Mu, K., Rossi, L., Sun, K., et al.: The effect of travel restrictions on the spread of the 2019 novel coronavirus (covid-19) outbreak. Science **368**(6489), 395–400 (2020)
14. Confederação Nacional do Transporte (CNT): Anuário CNT do Transporte 2022 (in Portuguese) (2022). Available at: https://anuariodotransporte.cnt.org.br/2022/Inicial. Accessed 25 October 2024
15. Departamento Nacional de Infraestrutura de Transportes: Nomenclatura das rodovias federais (2020). https://www.gov.br/dnit/pt-br/rodovias/rodovias-federais/nomeclatura-das-rodovias-federais/nomeclatura-das-rodovias-federais-1. Accessed 29 Oct 2024
16. Federal Republic of Brazil: Constituição da República Federativa do Brasil de 1988 (1988). Available at: https://www25.senado.leg.br/web/atividade/legislacao/constituicao-federal. Accessed 25 Oct 2024

17. Ferrari, C., Bottasso, A., Conti, M., Tei, A.: Economic Role of Transport Infrastructure: Theory and Models. Elsevier (2018)
18. Gastner, M.T., Newman, M.E.: The spatial structure of networks. Eur. Phys. J. B Condens. Matter Complex Syst. **49**, 247–252 (2006)
19. Guimera, R., Mossa, S., Turtschi, A., Amaral, L.N.: The worldwide air transportation network: anomalous centrality, community structure, and cities' global roles. Proc. Natl. Acad. Sci. **102**(22), 7794–7799 (2005)
20. Harvey, D.: The condition of postmodernity. In: The New Social Theory Reader, pp. 235–242. Routledge (2020)
21. Hulme, P.E.: Trade, transport and trouble: managing invasive species pathways in an era of globalization. J. Appl. Ecol. **46**(1), 10–18 (2009)
22. Kaluza, P., Kölzsch, A., Gastner, M.T., Blasius, B.: The complex network of global cargo ship movements. J. R. Soc. Interface **7**(48), 1093–1103 (2010)
23. Li, Z., Dawood, S.R.S.: World city network in china: a network analysis of air transportation network. Mod. Appl. Sci. **10**(10), 213 (2016)
24. Li, Z., Wu, M., Chen, B.R.: Is road infrastructure investment in china excessive? Evidence from productivity of firms. Reg. Sci. Urban Econ. **65**, 116–126 (2017)
25. Liu, C.M., Li, J.W.: Small-world and the growing properties of the Chinese railway network. Front. Phys. China **2**, 364–367 (2007)
26. National Department of Transport Infrastructure (DNIT): National road system database (snv) (2024). https://www.gov.br/dnit/pt-br/snv. Ministry of Transport
27. Newman, M.E.: The structure and function of complex networks. SIAM Rev. **45**(2), 167–256 (2003)
28. Newman, M.E.: Analysis of weighted networks. Phys. Rev. E Stat. Nonlinear Soft Matter Phys. **70**(5), 056,131 (2004)
29. Rodrigue, J.P.: The Geography of Transport Systems. Routledge (2020)
30. Sen, P., Dasgupta, S., Chatterjee, A., Sreeram, P., Mukherjee, G., Manna, S.: Small-world properties of the Indian railway network. Phys. Rev. E **67**(3), 036,106 (2003)
31. Skidmore, T.E.: Brazil: five centuries of change. OUP Catalogue (2009)
32. Soo, K.T.: Zipf's law for cities: a cross-country investigation. Reg. Sci. Urban Econ. **35**(3), 239–263 (2005)
33. Tak, S., Kim, S., Byon, Y.J., Lee, D., Yeo, H.: Measuring health of highway network configuration against dynamic origin-destination demand network using weighted complex network analysis. PLoS One **13**(11), e0206,538 (2018)
34. Weber, J.: The evolving interstate highway system and the changing geography of the United States. J. Transp. Geogr. **25**, 70–86 (2012)

Improving Flocking Behaviors in Street Networks with Vision

Guillaume Moinard and Matthieu Latapy

Abstract We improve a flocking model on street networks introduced in a previous paper. We expand the field of vision of walkers, making the model more realistic. Under such conditions, we obtain groups of walkers whose gathering times and robustness to break ups are better than previous results. We explain such improvements because the alignment rule with vision guaranties walkers do not split into divergent directions at intersections anymore, and because the attraction rule with vision gathers distant groups. This paves the way to a better understanding of events where walkers have collective decentralized goals, like protests.

Keywords Street networks · Flocking · Vision · Robustness · Protests

1 Introduction

Consider the following scenario. Protesters are scattered throughout a city and share the common objective to gather into groups large enough to perform significant actions. They face forces that may break up groups, block some places or streets and seize any communication devices protesters may be carrying. As a consequence, protesters only have access to local information on people and streets around them. Furthermore, formed protester groups must keep moving to avoid containment by adversary forces.

In this scenario, protesters need a distributed and as simple as possible protocol, that utilises local information exclusively and ensures a *flocking* behavior, i.e., the rapid formation of significantly large, mobile, and robust groups.

In a previous paper [1], the city was modeled as a network of streets and intersections, and protesters were biased random walkers on this network. Authors then identified some key building block for those protocols to guaranty walkers will gather in a short range of time, such as the *alignment* rule we will present in Sect. 3.2.

G. Moinard (✉) · M. Latapy
Sorbonne Université, CNRS, LIP6, Paris, France
e-mail: guillaume.moinard@lip6.fr

M. Macedo et al. (eds.), *Complex Networks XVI*, Springer Proceedings in Complexity,
https://doi.org/10.1007/978-3-031-93619-7_10

In this paper, while building on the same rules, we explore the effect of expanding walkers range of vision. Instead of applying their decision rules only to neighbor nodes in the graph, they will apply it to a set of nodes in a given range. We want to see if accessing more information in that manner improves walkers decision and enhances their flocking behavior.

In Sect. 2, we compare our approach to related works on flocking and gathering on networks. In Sect. 3, we present the framework of street networks and walking rules introduced in [1]. We then define walkers vision in Sect. 4 and introduce new rules we construct with this vision. In Sect. 5, we run extensive experiments to measure the effect of vision on our walking rules and their combination. We also explore the robustness of groups, measuring how they reform if adversary forces break them up while following an effective tactic. Finally, we summarise our contributions in Sect. 6.

2 State of the Art

The most famous flocking model the Reynold's model [2]. Since its publication, many papers studied flocking behaviors in a wide variety of contexts. However, most studies apply to continuous spaces, such as 2D or 3D spaces [3–6]. In this paper, we focus on street networks, and such graphs are discrete spaces.

Nevertheless, some articles have explored algorithms for gathering on any connected graph [7]. However, they require long term memory and computing capacities that exceeds what real pedestrians are capable of. Moreover, gathering is only a part of the flocking problem, as we also require walkers to subsequently move together once gathered.

Other articles have adapted rules proposed in the Reynold's model on a line of nodes [8]. With respect to our article, this case corresponds to the situation of walkers in a single street. However, we want to have well-defined behaviors on the entire network, which includes the intersections.

In [1], the authors presented a flocking model on street networks. Efficient behaviors emerged from a combination of multiple rules. We propose here a simpler and more realistic model inspired by Reynold's model, suited for any kind of network. Considering the importance of vision in the success of flocking models [9], we introduce a vision range for walkers and study the impact of its depth on flocking.

3 Framework

We need a framework to simulate displacements of protesters in a city. We model cities as undirected graphs we call street networks. Protesters are then biased random walkers on this network. They can move from node to node with simple rules we introduce in the following Section.

3.1 Discretized Street Networks

In order to model real-world cities, we leverage OpenStreetMap data and the OSMnx library [10]. For a given city, we use this library with its default settings to extract the graph $G = (V, E)$ defined as follows: the nodes in V represent street intersections in this city and the links in $E \subseteq V \times V$ represent pieces of streets between them. We take the undirected graph G, meaning there is no distinction between (u, v) and (v, u) in E. In addition, we denote by $N(v) = \{u, (u, v) \in E\}$ the set of neighbors of any node v in V.

Like [1], in the following, we use a typical instance, namely Paris, to present our work in this paper. This street network has 9,602 nodes, 14,974 links, leading to an average degree of 3.1. Its diameter is 83 hops and its average distance is 39.4 hops. The average street length is 99 m, and the average distance is 5,552 m.

The links of a street network generally represent street segments of very heterogeneous lengths [11]. Then, moves from a node to another one may have very different duration. In order to model this, we use the same discretization procedure as in [1]. It consists in splitting each link of the street network into pieces connected by evenly spaced nodes. We illustrate this procedure with our Fig. 1.

In the obtained graph, each link represents a street slice of length close to a parameter δ. Then a walker consistently make a move of length approximately δ at each hop.

Like in [1] we use δ equal to 10 m, leading to a network of $N = 130{,}276$ nodes and $M = 300{,}736$ links.

3.2 Walkers

Given a network $G = (V, E)$, we consider a set W of walkers numbered from 1 to $|W|$. We denote the location of walker i at time t by $x_i(t) = v$, with $v \in V$. We call *group* the set of walkers at a given node v at a given time t: $g_v(t) = \{i, x_i(t) = v\}$.

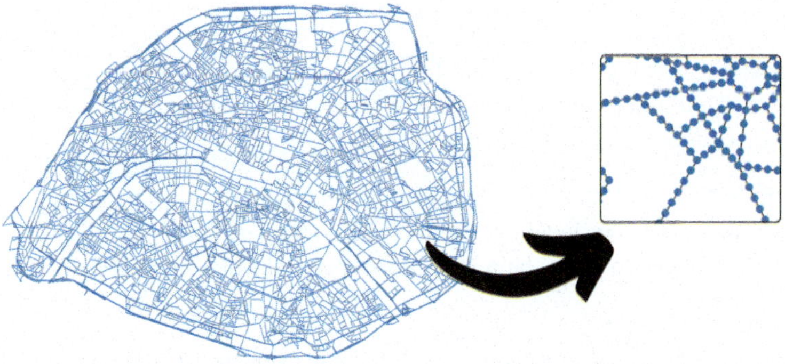

Fig. 1 A piece of the discretized street network around *Place de la Nation* in Paris

We denote by $g(t) = |\{g_v(t), v \in V, g_v(t) \neq \emptyset\}|$ the number of non-empty groups at step t.

We characterize *flocking* as a gathering of walkers exploring the network. In [1], big enough groups of walkers only form with rules that also make walkers explore the network. Therefore, the gathering of walkers is, on its own, a good indicator of the efficiency of our model.

Definition 1 (*Gathering score*) We denote the average size of non-empty groups of walkers by $\bar{n}(t) = \frac{W}{g(t)}$ and use this metric as our gathering score to evaluate the efficiency of our walks.

At each time step t, a walker i moves to a node $x_i(t+1) \in N(v)$. We present the two criteria, introduced in [1], a walker uses to determine its next location:

1. The *number of walkers* located at node u at time t; $n_u(t) = |g_u(t)|$
2. The *net flux* of walkers on a link $(u, v) \in E$ a walker perceives from node u; $J_{u \to v}(t)$ which is, at t, the difference between the number of walkers that entered the link from node u at $t-1$, and those who entered from node v.

Notice that the $J_{u \to v}(t)$ is a quantity that can be negative when the number of walkers that enter the link from v is greater than those who enter from u. Moreover, we always have $J_{u \to v}(t) = -J_{v \to u}(t)$, which implies the flux can be an attractive or repulsive force for walkers, whether they stand at u or v. We then can describe rules for the movement of walkers.

1. *Alignment rule*: A walker at node u at time t moves to the neighbor v that maximizes the net flux $J_{u \to v}(t)$.
2. *Attraction rule*: A walker at node u at time t moves to the neighbor v that maximizes the number of walkers $n_v(t)$.

So far those rules do not take into account the vision of walkers. We now present the vision-based rules that consider the depth of the vision of walkers.

4 Vision

We construct walkers vision in the following manner. We introduce the concepts of *vision depth* and *branches* to define how a walker combines the rules from the previous section with a wider field of vision.

Definition 2 (*Vision depth*) A walker can see what happens on the network up to a certain distance from its location, i.e., the vision depth d. This depth is an integer, the distance between two nodes being the number of links in a shortest path connecting them.

A vision depth equal to 1 means a walker i can evaluate the number of walkers on its neighbor nodes $v \in N(x_i(t))$ and the net flux on adjacent links $(x_i(t), v)$. It sees all information at distance 1 on the network. If this depth is greater than 1, a walker perceives the two criteria on nodes and links further than its neighborhood. On the other hand, the special case of a vision equal to 0 corresponds to a blind walker that can only follow a random walk.

Notice the depth of a walker vision can differ from one criterion to another. We call respectively d_n and d_j the depth of the vision for the number of walkers and for the net flux. A walker can see those quantities on nodes or links up to a distance d_n and d_j respectively.

Within this field of vision, a walker can look into different directions, at least one per neighbor nodes. We say a walker perceives multiple *branches*.

Definition 3 (*Branch*) First, a branch B is a simple path, i.e., a sequence of nodes $(b_0, b_1, \ldots, b_k)$ such that $\forall i \neq j, b_i \neq b_j$ and $\forall i, (b_i, b_{i+1}) \in E$. Moreover, a branch must not be the prefix of any other simple path. We denote by $\mathcal{B}(u)$ the set of all branches starting with $b_0 = u$.

A walker i considers all branches starting from its location $x_i(t)$. This means that, $\forall B \in \mathcal{B}(x_i(t))$, it evaluates the values of the two criteria respectively up to the d_n first nodes and d_j first links of the branch B.

On Fig. 2, we see the walker on node u with $d = 2$ considers the sequence of nodes in red, green, yellow and blue, as they are the d first nodes of the branches the walker can see. The walker can then evaluate the values of the two criteria on the nodes and links of the branch, and use them to decide its next move.

We notice that, in the set $\mathcal{B}(u)$, there can be multiple branches passing through each neighbor in $N(u)$. In a line of nodes, a single street, there is only one simple path in a given direction, and therefore the branch is unique. However, at intersections there are multiple branches. For example, in Fig. 2 we see the green and blue branches both start from node u but go into different directions after an intersection at node e.

Another comment concerns the length of branches with respect to the vision depth. We see that, in the case of the orange arrow, a branch can be shorter than $d = 2$, as

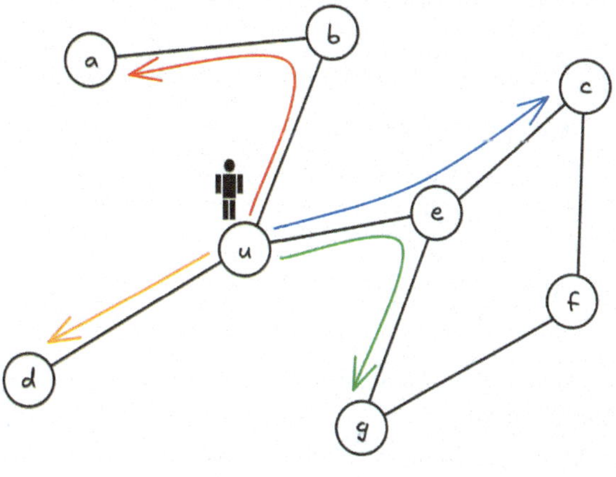

Fig. 2 Walker on node u with $d = 2$ has access to the information on the sequences of links and nodes highlighted by arrows, as they are the d first nodes and links of the branches in $\mathcal{B}(u)$

a simple path can not go back to a node already visited. The walker on u then only considers the nodes up to the end of the branch, i.e., u and d.

4.1 Weighting Branches

We know that $n_u(t)$ and $J_{v \to u}(t)$ are two commensurable quantities. Indeed, the number of walkers is a quantity that is directly comparable to the net flux, as a group of walkers all moving from u to v at $t-1$ produces a net flux so that $J_{u \to v}(t) = n_v(t)$. However, if their respective vision depths are different, we cannot compare the two criteria directly. We have to define a way to aggregate them, taking into account a walker might see more values from one criterion.

To do so, for a walker i, we denote by $w_B(d_n, d_j, t)$ the weight, at time t, of a branch $B \in \mathcal{B}(x_i(t))$ and define it as:

$$w_B(d_n, d_j, t) = \mathcal{N}_B(d_n, t) + \mathcal{J}_B(d_j, t) \tag{1}$$

where the two terms are the mean of each criterion values the walker considers along the branch:

$$\mathcal{N}_B(d_n, t) = \sum_{i=1}^{\Delta_n} \frac{n_{b_i}(t)}{\Delta_n} \qquad \text{with } \Delta_n = \min(|B|, d_n) \tag{2}$$

$$\mathcal{J}_B(d_j, t) = \sum_{i=0}^{\Delta_j} \frac{J_{b_i \to b_{i+1}}(t)}{\Delta_j} \qquad \text{with } \Delta_j = \min(|B|, d_j) - 1 \tag{3}$$

This guarantees we sum comparable quantities, as each term in $\mathcal{N}_B(d_n, t)$ and $\mathcal{J}_B(d_j, t)$ is divided by the number of terms in its sum. We can finally define a walker tactic as follows.

Definition 4 (*Tactic*) For a walker i at time t, following the tactic defined by the pair (d_n, d_j), is to identify the branch $B \in \mathcal{B}(x_i(t))$ with the maximum weight $w_B(d_n, d_j, t)$, and subsequently moves to the neighbour $v \in N(x_i(t))$ so that $v = b_1$.

Notice that, in the case of several branches with the same highest value, the walker picks one node randomly among them.

5 Experiments

In this section we measure how well our walkers perform at flocking. We seek to understand how the vision depth impacts their dynamics and if this new feature improves flocking behaviors. We run all experiments on the discretized Paris street network for 1000 steps with as many walkers as there are nodes in the network. They all start at random nodes.

To understand the impact of the vision depth on walkers behavior, we first measure the impact of the vision depth on the gathering score for each individual criterion. We will then allow walkers to make use of both criteria, with different vision depth for each of them, in order to identify the best combination.

5.1 Impact of Vision on Each Criterion

In this first experiment, walkers only make use of the attraction rule. In Fig. 3, we display the gathering score we obtain at the end of a run for each vision depth. It linearly increases with the vision depth. This is because walkers can see further and therefore interact with a number of walkers that is proportional to their vision depth. The gathering score is then roughly two times the vision depth, as a walker in a street sees exactly $2d_n$ nodes.

In the second experiment, walkers only make use of the net flux criterion. We know from [1] that this alignment rule, alone, creates a much richer process that does not simply converge after a few steps. That is why we see that, still on Fig. 3, this alignment rule is already better than the attraction one when walkers only see their neighbors ($d_n = d_j = 1$).

Moreover, its gathering score still increases with the vision depth. In [1] authors showed that, each rule taken separately, the net flux criterion is the most effective at

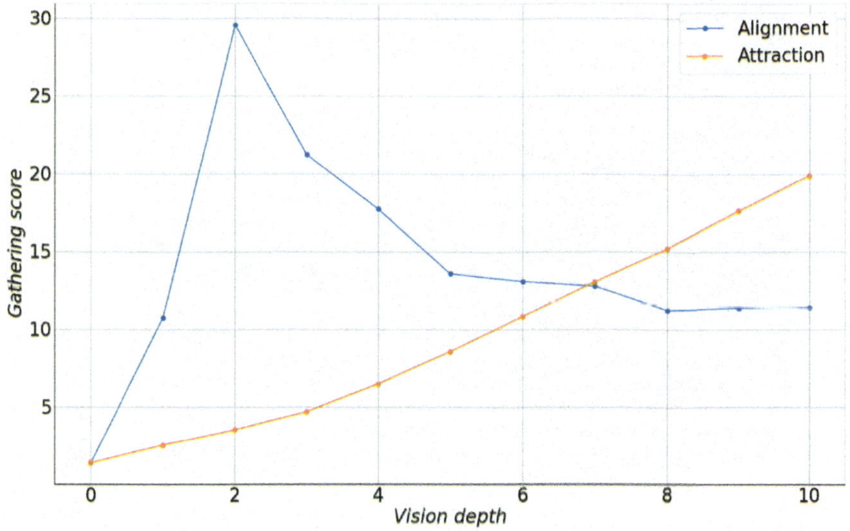

Fig. 3 Evolution of the last step gathering score for the alignment and attraction rules with different vision depths

gathering walkers, except that it can not prevent a group of walkers from splitting into two groups at an intersection. However, with $d_j = 2$, when a group of walkers splits at an intersection, the walkers that form the smallest new group now still perceive a positive net flux from the other walkers last move. They therefore come back and follow their original group. This means that, given $d_j = 2$, the alignment rule is able to have all walkers in the network flock together at long time.

However, for higher values of the vision depth, the gathering score decreases. This is because a walker can now see too far and therefore anticipates the arrival of other walkers from their net flux. In such conditions, a small group will align with a bigger in advance; its walkers will turn back and be repealed by the bigger group before the two merge. This results in groups of walkers avoiding each other, and therefore not gathering despite exploring the network.

5.2 *Impact of Vision on Combinations of Criteria*

We run the same experiment as before, with different vision depths for each criterion. We take only in count the gathering score at the last step of a run to see how different combinations of criteria play on flocking behavior. We display our results in Fig. 4.

Notice that the top most line and the left most column correspond to the experiments we presented in previous sections. With $d_n = d_j = 0$, the top left cell is the result for random walkers.

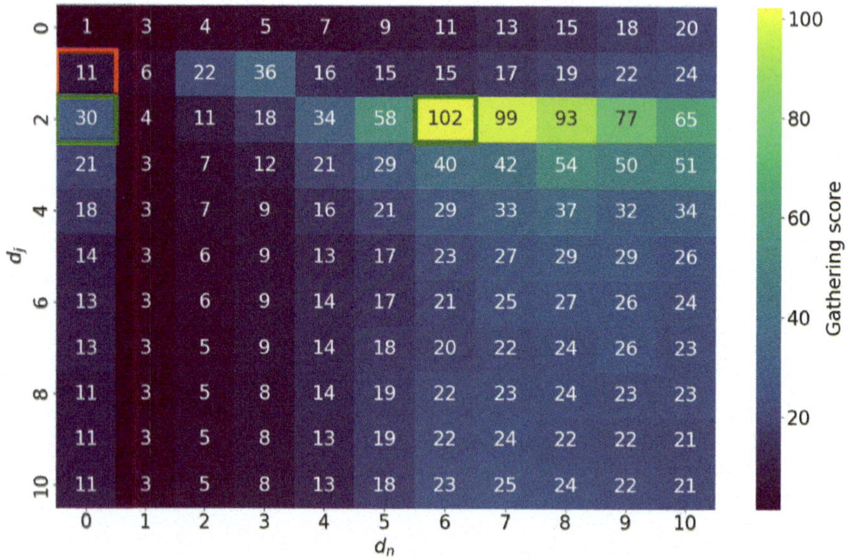

Fig. 4 Heatmap of gathering scores for every tactic with combinations of vision depths d_n and d_j from 0 to 10

First, the second column stands out from the previous and following columns, because of its poor scores. This is due to the vision depth for the attraction rule being equal to or smaller than the vision depth for the alignment rule. This implies that, while the alignment rule averages over multiple nodes, with some of them being empty, the attraction rule only takes into account the neighbor node. The contribution from the first term in (1) is then most of the time greater than the second term. Therefore, rather than slightly influencing a group that already flocks, the attraction rule becomes dominant. As this rule do not produce flocking on its own, the bigger the alignment vision depth, the more the score drops until being as bad as for the attraction rule alone.

This dynamic also explains why a value for at a cell (x, y) in the upper right triangle of the matrix is better than the one at (y, x) in the lower left one. Indeed, the upper right triangle corresponds to tactics for which the attraction vision depth for is greater than the alignment vision depth. In those cases, the attraction rule is able to influence the group that flocks, and the alignment rule is able to make the group flock. This is the best combination of vision depths for the two rules. This result is in line with the literature on flocking in continuous spaces [9], where rules often do not apply in the same neighborhood. More precisely the attraction rule generally applies to a further neighborhood than the alignment rule.

Notice the first column and row do not respect such a pattern, as they do not describe combination of the two rules. Moreover, in the bottom right corner, a few cells do not either, although the value gap between such cells and their symmetric cell in the matrix is very small.

In Fig. 4 we mark with a red square the cell showing the gathering score for walkers using the alignment rule with $d_j = 1$, like in [1]. We also highlight in green squares the two main contributions of this paper.

We first have the alignment rule with $d_j = 2$, that greatly increases groups size as we explained in Sect. 5.1. We now also see an optimal combination of the two rules, with vision depths $d_n = 6$ and $d_j = 2$. This combination takes advantage of the flocking behavior the alignment rules induces, and of the gathering effect from the attraction rule. This gives rise to groups that flock while being to move preferentially towards other groups. Such groups are up to 10 times larger than previous results. Increasing d_j obviously degrades the gathering score due to the repealing effect we highlighted. However, although it is not clear why, increasing d_n above 6 also worsen results. We also notice that, the further down the y-axis, the more the optimal combination on each line shifts to the right.

Finally, we display the evolution of the most interesting tactics in Fig. 5. All tactics using the attraction rule alone, whatever the value of d_n, are easy to describe; groups quickly form and do not evolve anymore, walkers being unable to move towards other groups that are now out of their range of vision. We notice that, in the first time steps and for big d_n value, walkers using attraction alone can form larger groups than those using any combination of the two rules. Therefore, if walkers only have a very limited amount of time to gather, the attraction rule alone is the best choice. However, if walkers have more time to gather, the combination of the two rules clearly becomes the best option.

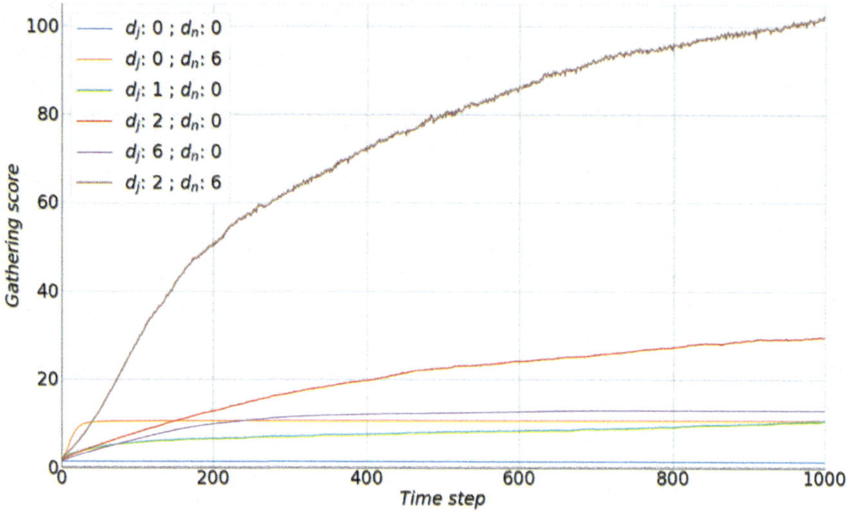

Fig. 5 Evolution of gathering score for some relevant tactics with different vision depths

5.3 *Robustness of Groups*

It is crucial for protesters to form groups that resist adversary forces that may break them up. In order to explore this robustness, we follow the method of [1] that models break ups as walkers suddenly following a random walk for one step. In this way, a group located at a given node splits into smaller groups that move to neighbor nodes, in a way similar to a group of protesters targeted by adversary forces.

More formally, we perform the following experiment: we run a simulation for 500 steps, then we impose walkers to move at random for one step. Finally, walkers use once again their combination of rules until the end of the run. We performed this experiment with the two tactics circled by the green squares in Fig. 4, i.e., the best tactic we found and the alignment rule with $d_j = 2$, and the tactic circled in red with $d_j = 1$.

Figure 6 displays the observed scores for the two latter, as the combination gave similar result to the alignment with $d_j = 2$, indicating that the attraction rule is not a key factor in the robustness of groups.

At the break up, groups size drops drastically. Then, for the alignment rule with $d_j = 1$, new groups are smaller than they would have been without the break up. This is a result already established in [1]. However, walkers following a tactic with $d_j = 2$ are able to recover from the break up and form groups even bigger than before. This is thanks to the longer vision depth that allows the walkers to interact further and therefore to recover from the break up.

In [1], combining multiple rules was enough to create robust groups that would recover from break ups. The gathering score would go back to its initial value as if the break up never happened. However, we show here that the vision depth is also

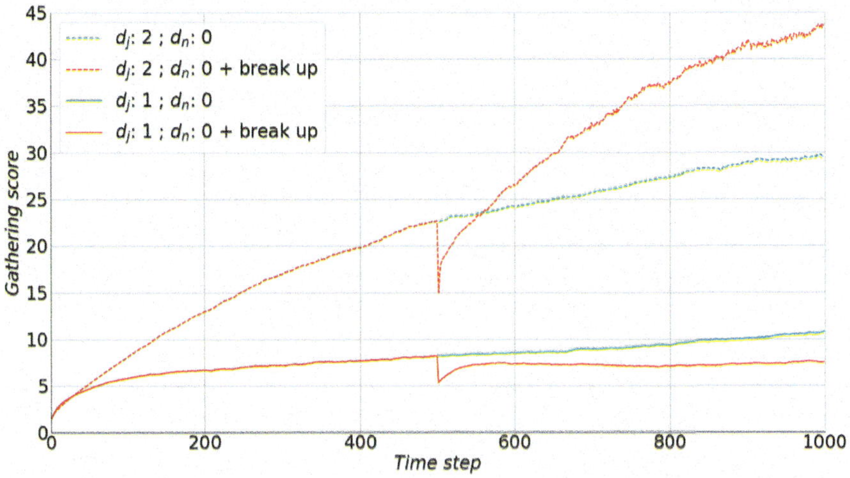

Fig. 6 Plot of robustness experiments, similar to Fig. 5. We display the gathering scores for tactics without break ups in green, with a break up in red

a key factor in the robustness of the groups that enables group to exhibit a form of anti-fragility, i.e., turning bigger after a break up.

6 Conclusion

In this paper, we presented a novel approach to produce flocking behaviors on a network which includes a parameter for the vision range of walkers. We studied our model on street networks, as they are a very relevant application case.

Doing so, we obtained better results than the original model in [1]. Indeed, expanding the vision of walkers for the alignment rule guarantees walkers do not split into multiple groups at intersection. Moreover, a tactic that combines attraction and alignment rules, with different vision depth for each, leads us to results ten times greater and such groups display remarkable robustness.

This simple model paves the way for studying other characteristics specific to flocking on a large variety of networks. Moreover, it can help to model and understand the behavior of pedestrians in urban environments.

Reproducibility. We provide an implementation of our models in C, and a Python code result analysis, with documentation, at: https://gitlab.com/guillaume_moinard/public-vision-flock.

Acknowledgements This work is funded in part by the CNRS MITI interdisciplinary programs.

References

1. Moinard, G., Latapy, M.: Fast flocking of protesters on street networks. In: IEEE/ACM International Conference on Advances in Social Networks Analysis and Mining (ASONAM) (2024)
2. Reynolds, C.W.: Flocks, herds and schools: a distributed behavioral model. In: Proceedings of the 14th Annual Conference on Computer Graphics and Interactive Techniques, pp. 25–34 (1987)
3. Vicsek, T., Czirók, A., Ben-Jacob, E., Cohen, I., Shochet, O.: Novel type of phase transition in a system of self-driven particles. Phys. Rev. Lett. **75**(6), 1226 (1995)
4. Cucker, F., Smale, S.: Emergent behavior in flocks. IEEE Trans. Autom. Control **52**(5), 852–862 (2007)
5. Olfati-Saber, R., Fax, J.A., Murray, R.M.: Consensus and cooperation in networked multi-agent systems. Proc. IEEE **95**(1), 215–233 (2007)
6. Dorigo, M., Theraulaz, G., Trianni, V.: Swarm robotics: past, present, and future [point of view]. Proc. IEEE **109**(7), 1152–1165 (2021)
7. Dessmark, A., Fraigniaud, P., Kowalski, D.R., Pelc, A.: Deterministic rendezvous in graphs. Algorithmica, **46**, 69–96 (2006)
8. Raymond, J.R., Evans, M.R.: Flocking regimes in a simple lattice model. Phys. Rev. E **73**(3), 036112 (2006)
9. Giardina, I.: Collective behavior in animal groups: theoretical models and empirical studies. HFSP J. **2**(4), 205–219 (2008)
10. Boeing, G.: OSMnx: new methods for acquiring, constructing, analyzing, and visualizing complex street networks. Comput. Environ. Urban Syst. **65**, 126–139 (2017)
11. Masucci, A.P., Smith, D., Crooks, A., Batty, M.: Random planar graphs and the London street network. Eur. Phys. J. B **71**, 259–271 (2009)

Social Circles Impact on Opinion Dynamics

Emy B. M. Bakker and Cristian Rodriguez Rivero

Abstract In this research, we contribute to increasing knowledge about how social circles impact opinion dynamics and we provide new insights on how opinion dynamics might contribute to automatically identifying social circles. To this end, we use the majority model as baseline model. We introduce a deterministic social circle approach to include social circle information into opinion dynamics simulation models. To evaluate the impact of social circles on the final distribution of opinions and the correlation between the final distribution of opinions and social circles, we introduce a metric of social consensus that translates the final level of agreement between individuals. We found that adding social circle information to opinion dynamics models results in a significantly lower final level of consensus than the baseline model which does not use social circle information. In addition, we conclude that there is a correlation between the final distribution of opinions and social circles. By comparing the mean level of consensus per social circle and per community, we could affirm that the correlation between the final distribution of opinions within social circles is higher than within communities. With this research we extend the knowledge of the role of communities in opinion dynamics to social circles.

Keywords Social network science · Opinion dynamics · Social circles · Majority model · Deterministic social circle approach · Level of consensus

---

E. B. M. Bakker (✉)
University of Amsterdam, Science Park, Amsterdam, The Netherlands
e-mail: emybakker996@gmail.com

C. R. Rivero
Department of Mining, Industrial and ICT Systems Engineering, Universitat Politecnica de Catalunya BarcelonaTech, Barcelona, Spain

Centre for Engineering Research in Intelligent Sensors and Systems (CeRISS), Cardiff Metropolitan University, Cardiff, Wales, UK

M. Macedo et al. (eds.), *Complex Networks XVI*, Springer Proceedings in Complexity,
https://doi.org/10.1007/978-3-031-93619-7_11

1 Introduction

As we are increasingly experiencing nowadays, online social networks are an important place for the dissemination of opinions. Dissemination is the spread of, for example, ideas, information, or news, and can result in consensus (everyone has the same opinion) or polarization (different opinions co-exist) due to network structures.

Structures in a social network include different (hierarchical) social circles. Hierarchical social circles can, for example, consist of individuals who participate in university, faculty, department, and cohort-based circles, which all represent different levels of interaction with different degrees of cohesiveness [12]. The question that results from this is which level of interaction matters the most to the process of opinion distribution. Once this is known, opinion dynamics in online social networks will become more comprehensible, allowing us to better understand the emergence of polarization and consensus [19]. At the same time, we can investigate the prospects of using opinion dynamics to automatically discover social circles.

Data extracted from online social media with well-defined social circles can help to find an answer to the formulated research question: *To what extent can social circles impact opinion dynamics?*

In this research, we investigate the impact of (hierarchical) social circles on opinion dynamics. For that, we use data on previously identified social circles, analyse it, and perform numerical simulations of opinion dynamics. Specifically, information about to which or to how many social circles a node belongs is used to provide a data-driven answer to the research question.

2 Social Networks

The foundation of social network science is graph theory in which the nodes (N) of a graph represent all individuals in the network and edges (also called ties or links) represent the connections between individuals. Visual examples of this are presented in Fig. 1. Hereafter, the words '*network*' and '*graph*' are used interchangeably.

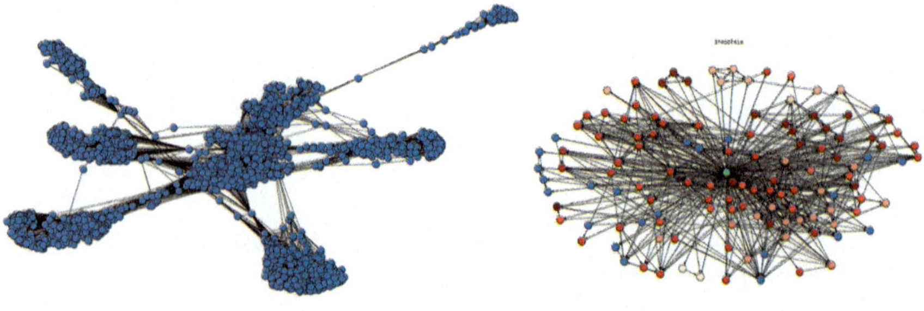

Fig. 1 **a** Facebook social network data set, **b** Twitter ego 374007416

Hermsen [10] and Otte and Rousseau [14] describe basic graph theory definitions as follows. The edges that are present in a network can be encoded in the form of an adjacency matrix $A \in \mathbb{R}^{N \times N}$. A non-zero value at A_{ij} expresses the presence of edge e_{ij}, so the connection between node i and node j. There are two kinds of networks, undirected and directed networks. For an undirected graph, it is not important what the direction of a connection is. In practice, this means that when two individuals are connected, they both know each other.

A (social) network can be partitioned into a community structure [7]. Communities can be seen as a group of nodes, joined together based on different possible conditions. In social networks this could represent a group of friends who share the same political beliefs, for example. It is expected that more edges exist within a community than between communities. Girvan and Newman [7] proposed a method for detecting such communities. They used edge betweenness (defined as the number of shortest paths between two nodes) to find community boundaries. Based on this method, Blondel et al. [3] came up with the commonly used Louvain method to detect communities. While classical algorithms tend to identify communities, an alternative network partition, more fine-gained, consist of social circles. Social circles, detailed in Sect. 2.2, can overlap between each other and represent groups of individuals with shared interests and/or backgrounds. These social circles can be identified in so-called ego networks [12]. An ego is an individual representing the focal node. The ego network is a local social network consisting of the connections between the acquaintances of the ego.

2.1 *Social Circles*

Social circles represent 'groups' in which an individual can categorize his or her network. This can be based on a variety of properties and groups can overlap with each other. To learn to identify social circles automatically, McAuley and Leskovec [12] created a machine learning model that incorporates user profile information into the network structure. They found that social circles tend to overlap heavily and are hierarchically nested. Generally, 25% of the circles contained completely within another circle, 50% overlapped with another circle, and 25% had no individuals in common with any other circle.

2.2 *Opinion Dynamics*

"Opinion dynamics are the processes that determine how opinions form and diffuse in society", as Menczer et al. [13] state. A widely used hypothesis about dynamics in social networks is the one of Granovetter [8]: "whatever is to be diffused can reach a larger number of people, and traverse a greater social distance, when passed through weak ties rather than strong". Weak ties refer to acquaintances that interact less frequently but do create a link between otherwise distant nodes (so with long ties). However tempting it is to simply respect this hypothesis, Centola and Macy [5] found that it is undesirable to just generalize it for every situation. In their research, they found that long ties are not always able to spread complex opinions (or as they call it, contagions) and can even impede dissemination completely. They defined complex contagions as contagions that require assertion from more than one acquaintance. The conclusion is that, for complex contagions, it can be beneficial to have a network with more clustering, even if the network as a whole has a larger diameter. The authors came to this conclusion by applying the small world principal model from Watts and Strogatz [20], which is able to randomly rewire a few local edges so that it reduces the mean distance between randomly chosen points. Based on [5], Centola [4] performed an online social network experiment and again found that individual adoption strongly improved for users who received assertion from more than one acquaintance. In addition, the spread went further and faster within a clustered network than in a random network. Furthermore, Arnaboldi et al. [1] estimated information diffusion in a social network based on the degree of trust between two individuals. They assume that dissemination primarily takes place on individuals that trust each other most. To analyse this, the authors estimate the degree of trust between individuals through the frequency of interaction they have and then use several thresholds to select the edges to be used for diffusion. They found that "active social contacts" (defined as individuals who communicate through at least one message per year) result in diffusion coverage of 96%. The more restrictive the contact threshold is, i.e. the more communication is needed, the less the coverage.

3 Methodology

In this research, we use data from three sources: Facebook, Google+, and Twitter. All data sets are open source and consist of network data with annotated ground-truth defined social circles [11]. Online social networks are a good representation of 'real-world' social networks, and it is not relatively easy to retrieve (network) data from online social networks [2, 17]. The Facebook data set is collected from survey participants using a Facebook app. The data set is anonymized in a way that profile information of different users can be matched on features, but it is not possible to see specific information about these features. For example, two users can match because they went to the same college, but it cannot be seen which college this is. For the Google+ data set, the collected data are from users that had manually shared their social circles using the 'share circle' possibility. The Twitter data set is retrieved from public sources. These data are based on user-defined lists, which are a way for users to organize their connections based on specific topics. All data sets include profile information (or so-called node features), ego networks, and social circles. To be able to analyze and visualize the data, we use the NetworkX package [9]. This is "a Python package for the creation, manipulation, and study of the structure,

Table 1 Data sets statistics

	Facebook	Google+	Twitter
Nodes	4039	107,614	81,306
Edges	88,234	12,674,353	1,768,149
Ego networks	7	132	973
Social circles	159	468	4065
Directed	No	Yes	Yes

dynamics, and functions of complex networks" [9]. The visualizations are color-coded for interpretation purposes: blue for the overall or 'neutral' level; (shades of) green for the level of ego networks; (shades of) red for the level of social circles; and purple for communities. More detailed statistics of the data sets are presented in Table 1.

3.1 Methods

3.1.1 Opinion Dynamics Modelling

As Sayama [15] and Menczer et al. [13] explain in their books, it is possible to "add states to nodes and dynamically update those iteratively". This process can be implemented and studied numerically to inform us about the long-term dynamics of opinions in real populations. In any type of model for influence diffusion or dynamics on networks, it can be assumed that a certain portion of the nodes is activated from the beginning. This can be encoded in the states of nodes. In the case of this research, it means that active nodes already 'accepted' the opinion and inactive nodes can be activated according to certain rules, conditions, and parameters. Those rules, conditions, and parameters depend on the type of simulation model. Models can be divided into discrete opinion models (individuals can have an integer number of opinions) and continuous opinion models (opinions can vary fluently from one extreme to another). In addition, models can also be divided into deterministic and stochastic models.

The results of deterministic models fully depend on the initial states and set parameter values. The results of stochastic models can be different each time—even when the initial states are the same—since there is always some level of randomness included. There are several simulation models for opinion dynamics on (social) networks that are commonly used. Generally, such simulation models take the following steps as described by Menczer et al. [13]. The code of Sayama [16] is used as source code. This code is adjusted to be able to implement social circle information and do the evaluations as discussed. For this research, the majority model is used as the baseline model. This model is commonly used in the field of opinion dynamics and can be classified as a discrete opinion dynamics model. Each individual changes their opinion to the opinion of the majority of its neighbors (also called the activation condition). When the case occurs where the number of neighbors with different opinions is equal, the opinion is updated with an equal probability for each opinion [13].

To introduce the impact of social circles on opinion dynamics in the model, two approaches are devised. The deterministic approach and the stochastic approach.

The deterministic social circle (DSC) approach. The deterministic approach can be seen as an extension of the majority model in combination with the well-known threshold model [13]. For this approach, the threshold to change opinion is still based on the opinion of the majority of individuals' neighbors. The difference, however, it that the contribution of each neighbor to achieving the threshold is weighted based on social circle information. In addition, this approach is still discrete in the sense that opinions are encoded in the states of nodes with either one or zero.

3.1.2 Evaluation Approach: Impact of Social Circles

In the social circle data as described in Sect. 3.1, only for the Google+ data set it is defined in the documentation that the ego nodes are included in their own social circles [11]. This is not visible in the data sets, so we add the ego node manually to its own social circles. For this research, it is assumed that the ego nodes are also included in their own social circles for the Facebook and Twitter data sets, so we add the ego nodes manually to its own social circles for those as well.

Average opinion. The average opinion can be used to analyze the results of the simulation models. This is the arithmetic average state over all nodes and can be created after every (amount of) iteration(s). Since the model starts with two opinions (either one or zero), the model will start with an average of around 0.5 as there is an equal probability for each opinion. In a steady-state, the average will become a specific value. If the steady-state concludes in consensus it will be less than 0.5 or more than 0.5, if it concludes in polarization it will be 0.5 [13]. We cannot say anything about impact with the final average opinion as a number on its own since the opinions are encoded with either one or zero. A final average opinion of 0.35, for example, would say that slightly more individuals end up with opinion 0, but it tells nothing about the level at which all individuals agree with each other.

Level of consensus. As a consequence of the shortcoming of the average opinion, we propose the level of consensus metric for this research. The level of consensus can be calculated by the average opinion minus a ratio of 0.5. By taking the absolute value of this times 2, we get a value between 0 (full polarization) and 1 (full consensus). The higher the level of consensus, the more consensus there is. Because we take the absolute value, the 'direction' (like the aforementioned example where slightly more individuals end up with opinion 0) does not matter. The metric is devised with the following formulation:

$$\text{global level of consequences} = 2\left|\frac{\sum_{i=1}^{n} o_i}{n} - 0.5\right| \quad (1)$$

where n is the number of all nodes and o_i is the opinion of node i. Finally, to verify whether the level of consensus differs between the baseline majority model and the DSC approach, we make use of the statistical t-test. This tests the hypothesis about two sets of data having significantly the same mean. To be able to use the Central Limit Theorem (with $n > 30$ we can assume the needed normality for the t-test) we run every simulation model 35 times [18]. Thus, to investigate the impact of social circles on the final distribution of opinions, we test if the mean of the level of consensus over 35 simulations differs between the majority model and the DSC approach.

3.1.3 Evaluation Approach: Correlation Between Final Distribution of Opinions and Social Circles

To say something about the correlation between the final distribution of opinions and social circles, we only consider the setting of non-weighted links. Social circle information for the weights is not used because the goal of this research is to give a contribution to increasing knowledge about how to automatically identify social circles by opinion dynamics. By using information about social circles beforehand, we would defeat that purpose. To this end, only the results of the majority model simulations will be used. For the correlation between the final distribution of opinions and social circles, an adjusted version of (1) can be used. Instead of using all nodes N, we calculate the level of consensus per social circle:

$$\text{level of consequences in social circle } C_k = 2\left|\frac{\sum_{i=1}^{nC_k} o_i}{nC_k} - 0.5\right| \quad (2)$$

where nC_k is the number of all nodes in social circle C_k and o_i is the opinion of node i. In addition, we make use of the Louvain method (discussed in Sect. 2.1) to detect communities [3]. To be able to apply this method for the directed (Google+ and Twitter) networks, we use the *igraph* package [6] instead of the NetworkX package [9]. We do this because the Louvain method in combination with the NetworkX package is not applicable for directed networks. The idea of the igraph package is the same but it is implemented in the programming language C instead of Python. After applying the Louvain method, we get to know automatically detected communities for which we also calculate the level of consensus with (2). Only in this case is nC_k the number of all nodes in community C_k. To get a clear view of the correlation between the final distribution of opinions and social circles, we do two comparisons: (1) we compare the global mean level of consensus with the mean level of consensus per social circle; (2) we compare the mean level of consensus per social circle with the mean level of consensus per community. If the global mean level of consensus is lower than the mean level of consensus per social circle, it means that individuals within a social circle tend to end up sharing the same opinion. Similarly, if the mean level of consensus within social circles is higher than the mean level of consensus within communities, it means that a network partition at the level of social circles can better capture the correlation of opinions. This is expected to observe in a social network. Additionally, we can compare a plot for the distribution of the level of consensus per social circle with one for the distribution of the level of consensus per community.

4 Evaluation

A visualization of the initial opinion distribution of the Facebook social network is presented in Fig. 2a. The white nodes are assigned with one opinion and the black nodes with the other. This same network but then after one simulation with the majority model is shown in Fig. 2b. The average opinion developed shows that the model became in a steady state around step 14. This could have been approximately a point to stop the simulation. This simulation process is repeated 35 times for the majority model and the DSC approach, but also for different influence parameter ρ values (see (1)).

4.1 Impact of Social Circles

The level of consensus for the Facebook social network is plotted for each type of simulation and are presented in Figs. 3 and 4. Each line/color represents one simulation. The global mean final level of consensus over all 35 simulations for the baseline majority model is 0.2276 (Fig. 3). For the DSC approach simulation with $\rho = 1$ this is 0.143 (Fig. 4). Figure 4 shows that simulations for the DSC approach often result in oscillations between two average opinions, while this does not happen for the majority model. We left it for future work to investigate if there is a specific

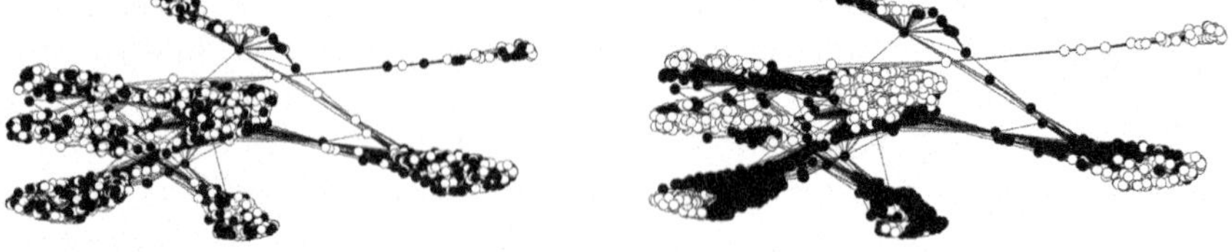

(a) (b)

Fig. 2 Facebook social network opinion distribution. **a** Beginning position. **b** After majority model simulation

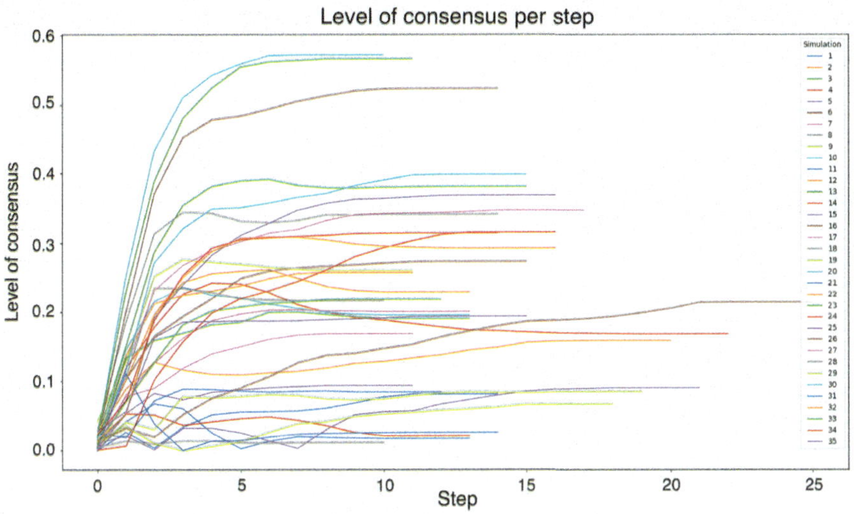

Fig. 3 Level of consensus per step for the Facebook social network majority model simulations

reason for this. With the t-test, we found a significant difference between the mean level of consensus of the majority model and the mean level of consensus of the DSC approach with $\rho = 1$. The p-value of the test is 0.01, which is smaller than the significance level $\alpha = 0.05$, so we use this as indication that the means are in all probability different. Different values of influence parameter ρ smaller and larger than 1 are tested to further research the impact of social circles. The mean level of consensus for Facebook of all simulations and their corresponding t-test p-values are given in Table 2.

We tested fewer values of influence parameter ρ for the Google+ social network. We did this in the first place because the Google+ social network is bigger than the Facebook social network and we found in the data exploration phase. In addition, from the results of the Facebook social network simulations it can already be concluded that adding social circle information to opinion dynamics models results in a lower final level of consensus. The mean final level of consensus over all 35 simulations for the majority model is 0.3315. For the DSC approach simulation with $\rho = 1$ this is 0.0392. The results of these simulations are given in Table 3.

For the Twitter social network, we did the same simulations as for different influence parameters ρ for the Twitter social network DSC approach (Table 4). The mean final level of consensus over all 35 simulations for the majority model is 0.2594. For the DSC approach simulation with $\rho = 1$ this is 0.0184. The results of these simulations are given in Table 5. From all the simulations in this section, we found that social circles do have impact on the final distribution of opinions. The height of the value of influence parameter ρ does not seem to have a relationship with the level of consensus. We left it for future work to research what this relationship could be, if any. However, for all the data sets we see that the simulations with the DSC

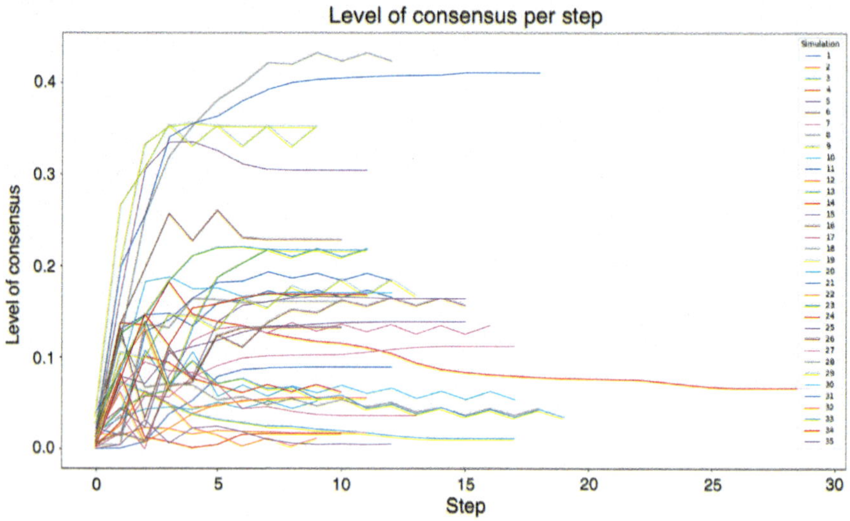

Fig. 4 Level of consensus per step for the Facebook social network DSC approach simulations with ρ = 1

Table 2 Data simulation results of different influence parameters ρ for the Facebook social network DSC approach

ρ	0.5	1	1.2	1.4	1.6	1.8	2
Mean level of consensus	0.1443	0.143	0.1561	0.1488	0.1357	0.1244	0.149
t-test *p*-value	0.0165	0.01	0.0207	0.01	0.0034	0.0008	0.0143

Table 3 Simulation results of different influence parameters ρ for the Google+ social network DSC approach

ρ	0.5	1	1.5	2
Mean level of consensus	0.0571	0.0392	0.0493	0.0533
t-test *p*-value	4.4227e−11	7.9004e−12	2.1036e−11	3.1713e−11

approach result in a significantly lower mean level of consensus than the simulations with the baseline majority model (all p-value < 0.05).

Therefore, we can conclude that adding social circle information to opinion dynamics models results in a lower final level of consensus.

Table 4 Simulation results of different influence parameters ρ for the Twitter social network DSC approach

ρ	0.5	1	1.5	2
Mean level of consensus	0.0163	0.0184	0.023	0.0162
t-test *p*-value	3.9831e−15	5.0774e−15	8.1755e−15	4.1196e−15

Table 5 Mean level of consensus after majority model simulations

	Facebook	Google+	Twitter
Nodes	0.2276	0.3315	0.2594
Edges	0.7873	0.5716	0.8166
Directed	0.7758	0.557	0.5402

4.2 Correlation Between Final Distribution of Opinions and Social Circles

As explained in Sect. 3.1.3, we did not consider social circle information to elaborate on the correlation between the final distribution of opinions and social circles. Therefore, the results in this section are only related to the majority model and are presented in Table 5. For all data sets, the mean level of consensus per social circle is higher than the global mean final level of consensus after 35 simulations. For the Facebook social network, the mean level of consensus increased with 0.5597; for the Google+ social network, it increased with 0.2401; and for the Twitter social network, it increased with 0.5572. From this, we can conclude that there is a correlation between the final distribution of opinions and social circles. This gives a good suggestion for future work in automatically identifying social circles by opinion dynamics.

For the Facebook social network, fifteen communities were detected with the Louvain 298 method; for the Google+ social network, 49 communities were detected; and for the Twitter social network, 87 communities were detected. If we compare the mean level of consensus per community with the mean level of consensus per social circle, an indication of a slight increase can be seen for the Facebook (mean level of consensus increased with 0.0115) and the Google+ (increase of 0.0146) social network. However, for the Twitter social network, a larger increase is observed (mean level of consensus increased with 0.2764). From this, we can affirm that the correlation between the final distribution of opinions within social circles is higher than the final distribution of opinions within (automatically detected) communities. The difference in how large this increase is between the different data sets will also be an interesting angle for future work.

5 Discussion

For this research, we used data from three sources: Facebook, Google+, and Twitter. The data sets include different types of social networks, small networks and bigger networks, a variety in the number of social circles, and undirected and directed networks. All opinion dynamics simulations are done on the three data sets, so the conclusions of this research are based on different types of online social networks. Online social networks are a good representation of 'real-world' social networks, and it is relatively easy to retrieve (network) data from online social networks. Since research in the field of social network science has hardly investigated the impact of social circles in the past, we did not know what to expect from the results of this research beforehand. We are surprised by the interesting conclusions that can be drawn and we think the results are a steppingstone for future work.

6 Conclusions and Future Directions

Research in the field of social network science has hardly investigated the impact of social circles on opinion dynamics. This research aimed to answer the question: *To what extent can social circles impact opinion dynamics?* We have shown, in various ways, that social circles can impact opinion dynamics. We did this by first characterizing and visualizing the previously identified social circles after which we used the majority opinion dynamics model as the baseline model. With this simulation model, we have identified the case of opinion dynamics where no social circle information is available. To include social circle information into the simulation model, we introduced the deterministic social circle (DSC) approach. The DSC approach showed the impact of social circles on the final distribution of opinions.

To conclude, at the start of this research we had the goal to contribute to increasing knowledge about how social circles impact opinion dynamics and how to automatically identify social circles through opinion dynamics. With the results of this research, we have met this contribution goal. We found that if information flows preferentially between individuals belonging to the same social circles leads to higher levels of polarization. Furthermore, we found that individuals belonging to the same social circle tend to end up sharing the same opinions.

Future directions are focused on the impact of social circles on the final distribution of opinions, particularly on the final levels of consensus and polarization and the correlation between the final distribution of opinions and social circles?" is analyzed by only considering the case of opinion dynamics where no social circle information is available.

References

1. Arnaboldi, V., Gala, M.L., Passarella, A., Conti, M.: Information diffusion in distributed OSN: the impact of trusted relationships. Peer-to-Peer Netw. Appl. **9**(2016), 1195–1208 (2016)
2. Bastian, M., Heymann, S., Jacomy, M.: Gephi: an open source software for exploring and manipulating networks (2009). http://www.aaai.org/ocs/index.php/ICWSM/09/paper/view/154
3. Blondel, V.D., Guillaume, J.-L., Lambiotte, R., Lefebvre, E.: Fast unfolding of communities in large networks. J. Stat. Mech. Theory Exp. **10**, P10008 (2008). https://doi.org/10.1088/1742-5468/2008/10/p10008
4. Centola, D.: The spread of behavior in an online social network experiment. Science **329**(5996), 1194–1197 (2010)
5. Centola, D., Macy, M.: Complex contagions and the weakness of long ties. Am. J. Sociol. **113**(3), 702–734 (2007)
6. Csardi, G., Nepusz, T.: The igraph software package for complex network research. Int. J. Complex Syst. **1695** (2006). https://igraph.org
7. Girvan, M., Newman, M.E.J.: Community structure in social and biological networks. Proc. Natl. Acad. Sci. **99**(12), 7821–7826 (2002). https://doi.org/10.1073/pnas.122653799; arXiv: https://www.pnas.org/content/99/12/7821.full.pdf
8. Granovetter, M.S.: The strength of weak ties. Am. J. Sociol. **78**(6), 1360–1380 (1973). http://www.jstor.org/stable/2776392
9. Hagberg, A., Swart, P., Chult, D.S.: Exploring network structure, dynamics, and function using NetworkX. Technical Report, Los Alamos National Lab (LANL), Los Alamos, NM, United States (2008)
10. Hermsen, F.A.W.: End-to-end learning on multi-edge graphs with graph convolutional networks. Master Thesis (2019)
11. Leskovec, J., Krevl, A.: SNAP datasets: Stanford large network dataset collection (2014). http://snap.stanford.edu/data
12. McAuley, J.J., Leskovec, J.: Learning to discover social circles in ego networks. In: NIPS, vol. 2012, pp. 548–56. Citeseer (2012)
13. Menczer, F., Fortunato, S., Davis, C.A.: A First Course in Network Science. Cambridge University Press (2020). https://doi.org/10.1017/9781108653947
14. Otte, E., Rousseau, R.: Social network analysis: a powerful strategy, also for the information sciences. J. Inf. Sci. **28**(6) , 441–453 (2002). https://doi.org/10.1177/016555150202800601
15. Sayama, H.: Introduction to the modeling and analysis of complex systems. Libretexts (2020). https://math.libretexts.org/Bookshelves/Scientific_Computing_Simulations_and_Modeling/Book:_Introduction_to_the_Modeling_and_Analysis_of_Complex_Systems_(Sayama)
16. Sayama, H.: PyCX. ver. 1.1 (2020). https://github.com/hsayama/PyCX
17. Stern, S., Livan, G.: The impact of noise and topology on opinion dynamics in social networks. R. Soc. Open Sci. **8**(4), 201943 (2021). https://doi.org/10.1098/rsos.201943; arXiv: https://royalsocietypublishing.org/doi/pdf/10.1098/rsos.201943
18. Triola, M.F.: Elementary Statistics, 12th edn. Pearson (2014)
19. Vasconcelos, V.V., Levin, S.A., Pinheiro, F.L.: Consensus and polarization in competing complex contagion processes. J. R. Soc. Interface **16**(155), 20190196 (2019). https://doi.org/10.1098/rsif.2019.0196; arXiv:https://royalsocietypublishing.org/doi/pdf/10.1098/rsif.2019.0196
20. Watts, D.J., Strogatz, S.H.: Collective dynamics of 'small-world' networks. Nature **393**(6684), 440–442 (1998)

The manufacturer's authorised representative in the EU is Springer Nature Customer Service Centre GmbH, Europaplatz 3, 69115 Heidelberg, Germany. If you have any concerns regarding our products, please contact ProductSafety@springernature.com

Printed and bound by CPI Group (UK) Ltd, Croydon, CR0 4YY
30/09/2026
02225469-0005